Current Topics in Tropical Cyclone Research

Edited by Anthony Lupo

Published in London, United Kingdom

IntechOpen

Supporting open minds since 2005

Current Topics in Tropical Cyclone Research
http://dx.doi.org/10.5772/intechopen.82936
Edited by Anthony Lupo

Contributors
Natalia Fedorova, Vladimir Levit, Lucas Cavalcante, Bo-Wen Shen, Tiffany Reyes, Boris Yurchak, Zhiyuan Wu, Mack Conde, Sullyandro Guimaraes, Zhong Liu, Chung-Lin Shie, Angela Li, David Meyer, Anthony Rocco Lupo, Brendan Heaven, Jack Matzen, Jordan Rabinowitz

First published in London, United Kingdom, 2020 by IntechOpen
IntechOpen is the global imprint of INTECHOPEN LIMITED, registered in England and Wales, registration number: 11086078, 5 Princes Gate Court, London, SW7 2QJ, United Kingdom
Printed in Croatia

British Library Cataloguing-in-Publication Data
A catalogue record for this book is available from the British Library

Additional hard and PDF copies can be obtained from orders@intechopen.com

Current Topics in Tropical Cyclone Research
Edited by Anthony Lupo
p. cm.
Print ISBN 978-1-83880-361-2
Online ISBN 978-1-83880-362-9
eBook (PDF) ISBN 978-1-83969-116-4

For EU product safety concerns:
IN TECH d.o.o., Prolaz Marije Krucifikse Kozulić 3, 51000 Rijeka, Croatia, info@intechopen.com or visit our website at intechopen.com.

Meet the editor

Dr. Anthony R. Lupo is an Atmospheric Sciences Professor in the School of Natural Resources at the University of Missouri. He earned his BSc degree from the State University of New York at Oswego in 1988, and his M.S. and Ph.D. degrees from Purdue University in 1991 and 1995. His research is in large-scale atmospheric dynamics, climate dynamics, and climate change, and he has more than 130 peer-reviewed publications. Additionally, he edited the books; *Recent Hurricane Research: Climate, Dynamics, and Societal Impacts* (2011), and *Recent Developments in Tropical Cyclone Dynamics, Prediction, and Detection* (2016) with IntechOpen. He is a member of the American Meteorological Society, National Weather Association, Royal Meteorological Society (Fellow), American Geophysical Union, Sigma Xi, Gamma Sigma Delta, Phi Kappa Phi, and Missouri Academy of Science (Fellow). He was a Fulbright Scholar to Russia (summer 2004) at the Russian Academy of Sciences in Moscow. He has won Fulbright scholarships to Belgorod State National Research University in Russia for 2014–2015, and fall 2017.

Contents

Preface

Since the start of the 21st century, changes in the current and future climate have been of greatest concern in the atmospheric and climate science communities. This includes changes in the frequency and strength of dangerous phenomena such as tropical cyclones. Although much research has been dedicated to this topic, there continues to be investigations into the dynamics of tropical cyclones including the dynamics that govern their movement. Remote sensing techniques and numerical models continue to be key tools in these investigations. In spite of all the advances in tropical storm research, these events continue to behave in ways that surprise us as the last few years have demonstrated.

While there will always be those events that develop rapidly, some storms have developed and then, sometimes unexpectedly, stalled over areas for a few days due to the lack of strong forces to steer them. This has led to some cities or regions being exposed to days of record setting rains or being lashed by strong winds for 24 to 48 hours. Heavy rain and flooding continue to be the most economically damaging and disruptive features associated with land falling tropical cyclones. There is research suggesting that in a warmer world, tropical cyclones will produce more precipitation since warmer air will support larger amounts of water vapor.

Therefore it is important to compile and publish the latest research every so often on recent developments in the area of tropical cyclone studies. This book includes several contributions of importance. This work will highlight tropical cyclone activity across the globe over the last decade and put this activity in the context of the previous 30 years. For the first time, all the major ocean basins have four relatively complete decades of tropical cyclone frequencies and good estimates of intensity. This work will show that the activity of the most recent decade is mostly similar to the decade before. Globally, the number of tropical cyclones has increased over the last four decades, but these increases are not statistically significant. Additionally, the increases were uneven across each ocean basin, and some such as the south Pacific have shown decreases.

In atmospheric science, there are several observational and model datasets that can be used in the study of tropical cyclones. The observational datasets are augmented by remotely sensed data. This work examines a couple of the most commonly used datasets and examines the precipitation and hydrology as well as the atmospheric dynamics. One study (Liu et al.) will demonstrate, using case studies, that different results will be obtained using datasets that may assimilate information from different platforms. For example, some of these datasets can produce a wet bias in their results. This study describes three main datasets and gives details with respect to data access, documentation, and user services as well. Another study (Guimarães) evaluates models that are part of the Coupled Model Intercomparison Project Phase 6 (CMIP6) and using the quantity Accumulated Cyclone Energy (ACE) with respect to recent climate and what the results mean to the future. This study compared model performance to the current observations, and then examined the future. There was no obvious trend for future scenarios of ACE, and no clear majority of increases or decreases of this quantity by different modeling groups.

While many studies examine the synoptic and dynamic features associated with the growth of tropical cyclones or their landfall, relatively few (Fedorova et al.) examine several case studies of weaker storms near the equator that give rise to fog and mist events over Northeast Brazil. These events can have serious consequences for aerospace operations. In these cases, it was found that the tropical cyclones or disturbances do not come ashore directly, but it is the indirect influence caused by secondary circulations that lead to the formation of fog and mist events. Also, tropical cyclones such as tropical storms and hurricanes generally emerge from tropical disturbances. Over the Atlantic region these disturbances occur with some regularity emanating from formation regions over Africa (African Easterly Waves – AEWs). Reyes and Shen use this quasi-periodicity to develop algorithms to forecast their occurrence using recurrence theory. These results show some promise in extending the lead time for the forecasting of the tropical cyclone "seeds".

Other parts of the world, such as the countries of the Northwest Pacific, are also vulnerable to the occurrence of tropical cyclones. A review by Wu and Conde summarizes the previous research done by the authors and collaborators involving the response of coastal ocean dynamics to tropical cyclone occurrence for variables such as sea surface temperature, sea surface salinity, storm surge simulation, and extreme rainfall. They also propose three research paths for future work that involves observational, dynamic, and model studies in order to improve the understanding of the future threat these regions may face due to tropical cyclones.

Satellite techniques are often used to determine the intensity of tropical cyclones using the area of cold cloud tops near the center as well as storm feature shapes as determined using visible or infrared imagery (the Dvorak technique). The work of Yurchak proposes using the character of the spiral rainbands as derived from not only satellite techniques, but augmenting this with ground and aircraft-based RADAR to determine the shape of the spiral bands. This information can be used to develop a hyperbolic-logarithmic streamline equation that ultimately relates the model to maximum wind speed.

Lastly, it was a pleasure to edit this book and work with all the authors and the staff of IntechOpen. The process was rather smooth. I am grateful to all the authors for their contributions, and the staff at IntechOpen for their hard work in keeping the process moving along. This is the third such project that I've been involved with and the experience has been good each time. Finally, it is my hope that the tropical meteorology community finds this book to be a useful resource for augmenting their own studies as well as providing a basis for future research.

Anthony R. Lupo
Atmospheric Sciences Program,
School of Natural Resources,
302 Anheuser Busch Natural Resources Building,
University of Missouri,
Columbia, MO, USA

Section 1

Climatological Behavior

Chapter 1

The Interannual and Interdecadal Variability in Tropical Cyclone Activity: A Decade of Changes in the Climatological Character

Anthony R. Lupo, Brendan Heaven, Jack Matzen and Jordan Rabinowitz

Abstract

During the last decade, there has been concern that the frequency or intensity of tropical cyclones (TCs) has increased. Also, climate models have shown varying results regarding the future occurrence and intensities of TC. Previous research from this group showed there is significant interannual and interdecadal variability in TC occurrence and intensity for some tropical ocean basins and sub-basins. This work examines global TC occurrence and intensity from 2010 to 2019 and compares this period to the same quantities from 1980 to 2009. The data used here are obtained from publicly available TC archives. Globally, the number of TC occurring over the latest decade is similar to the previous decade. However, while the 40-year trend shows an increase in TC, only intense hurricanes have shown an increase. The Atlantic Ocean and North Indian Ocean Basins show increases in TC activity, especially intense storms. The Southern Hemisphere and West Pacific Region show decreases in TC activity. In the West Pacific, intense TC did not increase, but the fraction of storms classified as intense increased. Only East Pacific TC activity showed no significant short- or long-term trends. Interannual and interdecadal variability in each sub-basin was found and there were some differences with previous work.

Keywords: tropical cyclones, climate change, variability, ENSO, intensity, PDO

1. Introduction

A decade ago, Lupo [1] found no statistically significant long-term trends in global tropical cyclone (TC) activity or in many of the regional basins, although detailed records for some parts of the globe (e.g., the Southern Hemisphere) have only been available since about 1980. This study looked at time series of varying lengths within each ocean basin. This same work showed that there was interannual variability in TC occurrences and intensities found in most ocean basins. However, there was little statistically significant interannual TC variability during the negative or cold phase of the Pacific Decadal Oscillation (PDO), but interannual TC variability with respect to El Nino and Southern Oscillation (ENSO) was enhanced during the positive or warm phase of the PDO. Lupo [1] also showed some interannual variability in the length of the TC season in different basins as well.

Others (e.g., [2], and references therein) found significant interdecadal TC variability in the Atlantic Region as related to teleconnections such as the North Atlantic Oscillation (NAO) or the Atlantic Multidecadal Oscillation (AMO), and relate these to a relative minimum in this region's TC activity in the late twentieth century and a sharp increase in TC activity for the early twenty-first century. These results were consistent with the results of Lupo [1]. Camargo et al. [3] examine the climatological character of TC including long- and short-range variability in each ocean basin as well. This work is a comprehensive review of those that relate TC activity to intraseasonal phenomena such as the Madden Julian Oscillation (MJO), ENSO, the Quasi-biennial Oscillation (QBO), and others.

Since Lupo [1], others have published results showing that there have been more recent increases in the number of stronger storms in both the Atlantic (e.g., [4, 5]) and the West Pacific basin [6]. The latter showed this trend has been occurring since 1998, but others have demonstrated that the trend has been present in the West Pacific since the 1970s (e.g., [7]). Globally, several studies (e.g., [8, 9]) have demonstrated an increase in the most intense storms and/or the associated precipitation rates [10]. The latest published study [11] examined the global frequency of intense TC from 1979 to 2017 and found statistically significant increases as well. Some have noted that these increases in intense TC are associated with basin-wide changes in the sea surface temperature patterns (e.g., [6]). Others (e.g., [12]) examined the rapid intensification of TC over the Atlantic Region during the latter part of the 20th century as related to climate variability and trends. Additionally, the IPCC [4] fifth assessment report demonstrated no general agreement about the relative contribution of natural and anthropogenic forcing to changes in TC intensity.

The focus on the most intense TC during the last decade is likely due to the fact that many climatological studies have established well the general character of TC climatologies in the world's ocean basins. Additionally, the contributing dynamics to TC formation, development, and decay are well known (e.g. [3, 13, 14], and references therein). At the turn of the twenty-first century, tropical cyclone (TC) activity and how this may change in the future were of great interest to the atmospheric science community (e.g., [15, 16]). Furthermore, there is interest in the observed and potential increase in rainfall rates [10]. Increases in intensity and rainfall rates could threaten vulnerable coastal areas.

The consensus of several global and regional scenarios for TC activity continues to project that the annual frequency will remain similar to today or decrease throughout the twenty-first century, but the intensity will increase (e.g., [4]). This may be due to projected decreases in strong tropical convection, although the confidence in this particular projection is lower. Additionally, these TC projections have been identified as uncertain since high-resolution simulations struggle to adequately capture TC occurrence and intensity [17, 18]. Also, the actual count of TCs is dependent on the different detection methods [19].

However, as discussed above with reference to Lupo [1] and further in that publication, the available time series across each region of the globe is uneven, and the ability to observe TC has continuously improved. There have even been changes in the instrumentation during satellite era and some studies (e.g., [20]) were able to homogenize the most recent satellite-derived data in order to analyze trends in TC occurrence and intensity. Also, changes in the techniques used to determine TC character and intensity have occurred as well (e.g., [21–23]).

The goal of this work is two-fold. The first is to examine the latest decade in TC activity in order to determine if there have been any major changes in global, regional, or subregional TC frequency since Lupo [1]. The activity of the latest decade will be placed into the context of previous activity going back to 1980 and recent studies where available. By going back to 1980, this work will present the occurrence and intensity of TC in every ocean basin (and sub-basin) where they occur over four decades and this work is one of only a few thus far (e.g., [11]). TC intensity was not available for all ocean basins until approximately 1980 (e.g., [1]). The techniques used here are the same as those found in Lupo [1] and earlier studies in order to facilitate comparisons to these older studies published by this group. The second will examine TC activity with respect to interdecadal variability, and in particular the PDO, in each region in order to determine whether the results of Lupo and Johnston [16] and Lupo [1] remained intact. While the examination of interannual and interdecadal variability of TC is not unique, the study of these quantities over the entire globe and in each TC basin and sub-basin for this recent 40-year period is the first as far as the authors are aware.

2. Data and methods

The data and methods are similar to those used in Lupo [1] and references therein, and more detail regarding some of these subjects can be found there. This study will examine all the globe's ocean basins and includes tropical storm occurrences as well. The global ocean basins (**Figure 1**) are as follows: the North Atlantic, East Pacific, West Pacific, North Indian, Southern Hemisphere (includes South Indian and the South Pacific), and the South Atlantic. Following Lupo and Johnston [16] and Lupo [1], the North Atlantic was divided into west and east along 45°W. The East Pacific is divided along 125°W and 20°N as in Collins [24, 25], while the West Pacific is divided up into 140°E and 20°N following Lupo [1]. The Indian Ocean in the Northern and Southern Hemisphere is divided into west and east along 75°E. The southwest and southeast Pacific are divided by 180° longitude. Both the Indian and SH sub-basin divisions followed Lupo [1]. TCs were assigned to the basin and sub-basin in which they first reached tropical storm status. A study of the background atmospheric and oceanic variables contributing to TC formation is not performed here as it is beyond the scope of this work.

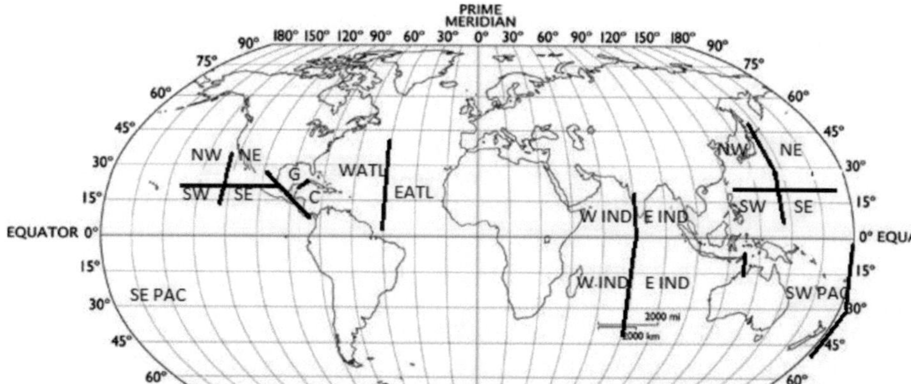

Figure 1.
The globe with the borders of each subregion for the North Atlantic (G = Gulf of Mexico, and C = Caribbean), East Pacific (125°W 20°N), West Pacific (140°E 20°N), northern Indian (75°E), and southern hemisphere (75°E in the Indian and 180° in the Pacific).

2.1 Data

The TC occurrence and intensity data for all basins since 1980 were downloaded via the UNISYS website (http://weather.unisys.com), although these data can also be found in the National Hurricane Center (NHC) or the Joint Typhoon Warning Center (JTWC) archives (e.g., [21]). The TC data since at least 1900 can be found in the International Best Track Archive for Climate Stewardship (IBTrACS) or the best track data archive [26, 27]. A description of these data sets and their reliability can be found in references, such as Landsea [28], Knapp et al. [26], or Kossin et al. [20]. Here, we use the term TC to include both hurricanes and tropical storms (TSs) following Lupo et al. [29] and references therein. TS refers only to those entities that obtained maximum wind speeds between 35 and 64 kt. The year 1980 was chosen for this study in order that time series of the same length can be compared across the globe since TCs were not categorized in the Southern Hemisphere until that year (see [1]). Also, TC data sets from before the satellite era may be missing TC occurrences that went undetected by ship or aircraft (e.g., [30]). Additionally, this study will compare and contrast briefly the most recent four decades with those previous to 1980 (see [16]) where those data exist (Atlantic Region and West Pacific Region). Hurricane intensity was rendered using the maximum wind speed attained during the lifetime of the storm. However, since wind speed data have relatively large measurement error, the Saffir-Simpson [31] hurricane intensity scale values were used here. In order to further eliminate problems with some of the data as discussed in Lupo and Johnston [16] and later studies, we combined hurricane intensity categories (Category 1 and 2—weak; Category 3, 4, and 5—intense) following Landsea [28].

2.2 Methodology

Arithmetic means and correlations were analyzed, and means were tested for statistical significance using a two-tailed z-score test, assuming the null hypothesis (e.g., [32, 33]). Intensity distributions were also tested using a χ^2 statistical test. These distributions were tested using the total sample climatology as the expected frequency and a subperiod as the observed frequency. The χ^2 test was used to test the intensity distributions (TS and Category 1–5) of the most current decade against those of the previous 30 years as well as to examine the interannual or interdecadal variability of intensities. It has been hypothesized that using the climatological frequency as the "expected" frequency is more appropriate than using an approximated distribution since such analytical distributions (e.g., Poisson distribution) may not adequately represent real-world distributions (e.g., [34]). It should be cautioned that while statistical significance reveals strong relationships between two variables, it does not imply cause and effect. Conversely, relationships that are found to be strong, but not statistically significant may still have underlying causes due to some atmospheric forcing process or mechanism (e.g., [34]). The long-term trends were tested for statistical significance using analysis of variance techniques (ANOVA) and in particular the F-test [32, 33].

The following descriptions can be found also in Lupo and Johnston [16]. The data were stratified by calendar year in the Northern Hemisphere (NH). In the Southern Hemisphere (SH), the tropical cyclone year is defined as the period beginning on July 1 and ending on June 30. For example, July 1, 2018 to July 1, 2019 was defined as 2019 since the majority of the SH TC season takes place from approximately December through April. We then analyzed the annual and monthly distributions of TC occurrence in order to find trends in TC season length or both the total sample and each intensity category.

2.3 Interannual and interdecadal variability

The Japan Meteorological Agency (JMA) El Nino and Southern Oscillation (ENSO) Index was used in this study. A list of El Niño (EN), La Niña (LN), and Neutral (NEU) years used here are shown in **Table 1**. A description of the JMA ENSO Index can be found on the Center for Oceanic and Atmospheric Prediction Studies website (http://coaps.fsu.edu/jma.shtml) hosted by Florida State University. In summary, this index is widely used (see [35]) and is defined by the long-term running mean sea surface temperature (SST) anomalies from the Niño 3 and 3.4 regions in the central and eastern tropical Pacific (e.g., [36]). The SST anomaly thresholds used to define EN years are those greater than +0.5°C, less than − = 0.5°C for LN years, and NEU otherwise. The JMA ENSO criterion defined the EN year as beginning on October 1 and ending on September 30. For example, ENSO year 1982 began on October 1, 1982 and ended on September 30, 1983. This definition, however, was modified here so that the EN year commenced with the initiation of the NH hurricane season (approximately June 1) following Lupo and Johnston [16] and used in Lupo [1]. This modification was necessary since EI Nino conditions typically begin to set in well before October 1, and the period August to October is close to the climatological peak of the hurricane season for the NH. No modification was needed for the SH. Additionally, while the JMA ENSO Index is more sensitive with the definition of LN than other indexes, it is less sensitive overall [37].

The Pacific Decadal Oscillation (PDO) is a 50- to 70-year oscillation described in the late twentieth century (e.g., [38, 39]) within the Pacific Ocean basin. We define the epochs of the PDO as found in Lupo et al. [35] and these are also cataloged at COAPS. The positive phase persisted from 1977 to 1998, while the negative phase has persisted since 1999. The most recent negative phase encompasses the most recent two decades, while the decades of the 1980s and 1990s are largely characterized by the positive phase. Where the data exist before 1980 (the Atlantic and western Pacific Regions), we can use the results of Lupo and Johnston [16] to characterize the negative PDO years from 1947 to 1976.

An in-depth discussion is found in Lupo [1] describing why these two teleconnections were used primarily to define interannual and interdecadal variability, in spite of the fact that many studies (e.g., [2, 3], and references therein) have shown for example that variability in the Atlantic Ocean Basin can be linked to

EN	NEU	LN
1982	1979–1981	1988
1986–1987	1983–1985	1998–1999
1991	1989–1990	2007
1997	1992–1996	2010
2002	2000–2001	2017
2006	2003–2005	
2009	2008	
2014–2015	2011–2013	
2018	2016	
	2019	

Table 1.
The list of ENSO years as found in Lupo et al. [35] and references therein.

teleconnections there. While the NAO-related variations in TC activity can make interpretation of PDO-related hurricane variability more difficult, there is substantial overlap between the PDO and the interdecadal modes of the North Atlantic Oscillation (NAO) [35]. Nonetheless, ENSO is a main driver of interannual TC activity in many ocean basins as demonstrated by many studies (e.g., [3]), and since PDO can be shown to modulate ENSO behavior, the focus here will be on these teleconnections.

3. Global tropical cyclone activity from 2010 to 2019

3.1 A comparison of tropical cyclone activity since 1980

In Lupo [1], tropical cyclone activity was examined within each ocean basin over different time periods. Here tropical cyclone activity since 1980 only was examined for each ocean basin and globally (**Table A1**). Globally, there has been no statistically significant trend in overall TC activity over the last 40 years and this is consistent with recent studies (e.g., [4]) (**Tables A1f** and **2**, **Figure 2a** and **b**). There was also no significant difference in the TC intensity distributions when comparing those of the most recent decade versus the 1980–2009 period. A noteworthy change implied in the global data set was an increase in tropical storm activity since 2000 at the expense of weaker (Category 1 and 2) hurricanes. However, the most recent decade (2010–2019) did not show appreciable changes worldwide when compared with the previous decade (2000–2009) or with the 1980–2009 period. There was, however, a significant upward trend in the number of Category 3–5 and 4–5 storms significant at the 99% confidence level (**Table 2**) consistent with Elsner [6, 10], or Kossin et al. [11].

An examination of each ocean basin demonstrates that only the ATL (**Table A1a** and **Figure 2c** and **d**) and NIND (**Table A1d** and **Figure 2i** and **j**). Regions experienced statistically significant increases for the trend in hurricane activity (at the 95% and 99% confidence levels, respectively) and total TC activity (at the 99% confidence level in both regions) (**Table 2**). Both regions showed slightly more activity in the most recent decade (2010–2019). Testing the distribution of TC intensities in both regions showed no statistically significant difference between the distribution of these for the most recent decade versus 1980–2009 using the χ^2 test. However, the ATL increases are most notable in the tropical storm category (**Table A1a**), but with little change in the weak hurricanes. In the NIND Region, however, these increases were noteworthy only for the number of hurricanes, especially major hurricanes (Category 3 and 4). In both of these regions, the increase in the trend for major hurricanes categories was significant at the 99% confidence level. A comparison to Klotzbach and Gray [2] or Lupo [1] showed that the ATL Region trends found here are consistent with those found for the late twentieth or early twenty-first century identified in those publications. Thus, this region has a longer history of increasing activity. In Lupo [1], the NIND Region showed little trend in TC activity. The upward trends in all categories for the NIND noted here (**Table 2**) could be a real phenomenon or a function of better detection and classification.

The increases were offset by overall decreases in the WPAC (**Table A1c** and **Figure 2g** and **h**) and SHEMI Regions (**Table A1e** and **Figure 2l** and **k**), which would show decreases, but only the decrease in WPAC hurricanes and SHEMI total TC were significant at the 99% confidence level (**Table 2**). Both regions were less active in the most recent decade (2010–2019). In the WPAC (**Table A1c**), the results found here were complex but contradict the results cited in section one.

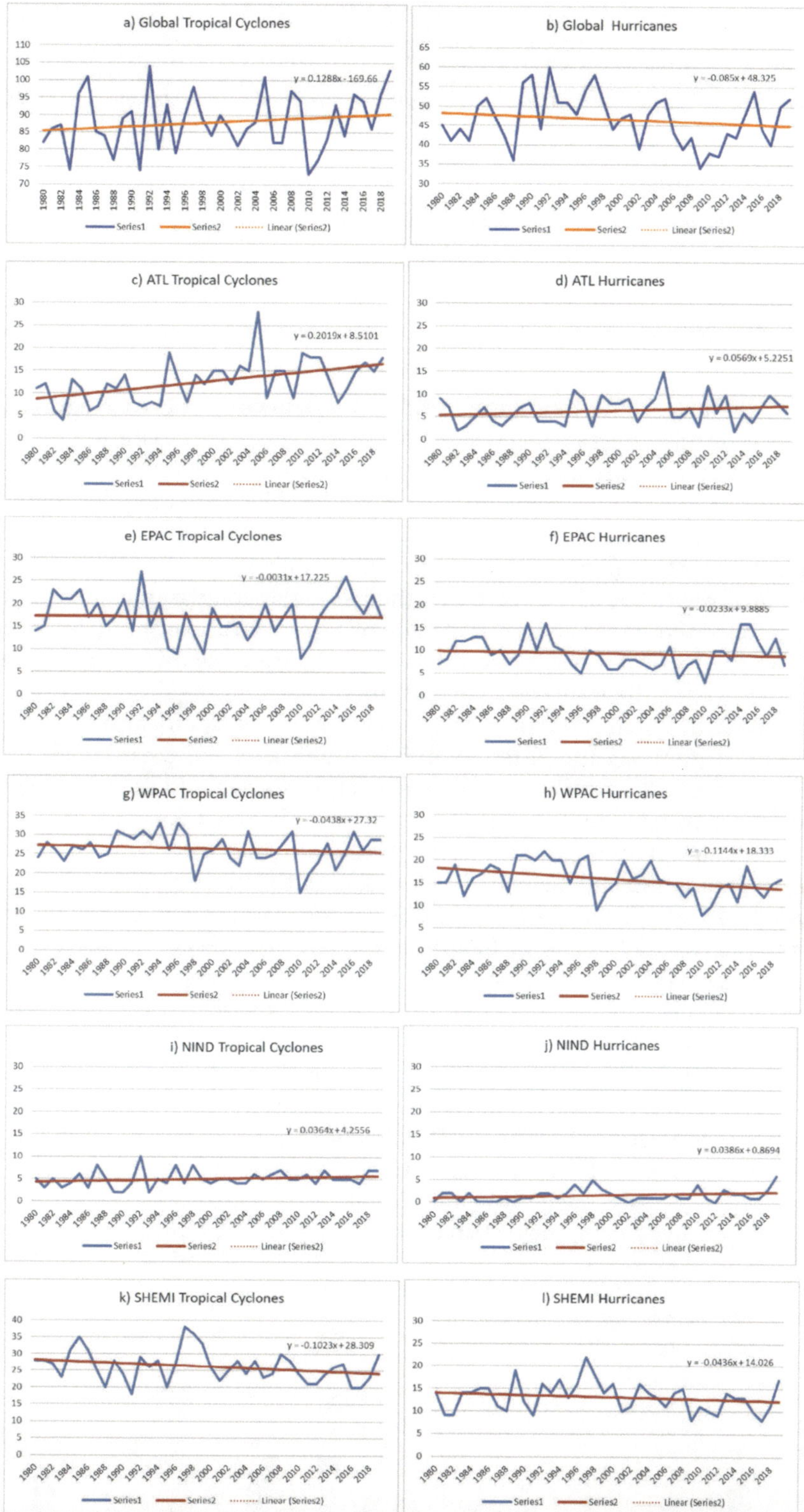

Figure 2.

The annual occurrence of (a) and (b) global, (c) and (d) Atlantic, (e) and (f) East Pacific, (g) and (h) West Pacific, (i) and (j) North Indian, and (k) and (l) Southern hemisphere tropical cyclones (left) hurricanes (right) from 1980 to 2019. The orange line is the linear trend line in each figure. The abscissa is years and the ordinate is annual occurrence.

	ATL	EPAC	WPAC	NIND	SHEMI	Globe
TS + Hur	**0.202****	−0.003	−0.044	**0.039****	**−0.102****	0.129
Tot Hur	**0.056***	−0.023	**−0.114****	**0.040****	−0.044	−0.085
Cat 3–5	**0.043***	0.007	−0.007	**0.029****	**0.044***	**0.117****
Cat4–5	**0.032***	0.034	0.030	**0.017****	**0.088****	**0.202****

*The value given is the slope of the trend line. Statistically significant values are bold, while those marked with a *, ** are significant at the 95%, 99% confidence level, respectively.*

Table 2.
A summary of the statistical significance for trends within each ocean basin.

Examining the major hurricanes, the trend was downward for the Category 3–5 TC, but upward for the Category 4–5 results. Neither trend was statistically significant (**Table 2**), and the significant downward trend was noteworthy in TC Category 1–2 (not shown). While this does not agree with studies like Zhao et al. [6], who have found an increase in intense TC over the WPAC, the decrease in weaker TC means that a greater percentage of WPAC TC was in the major category. During the past two decades, about 60% of TC were classified as major compared to 50% in the two decades prior to those (see also [1]). In spite of a greater ratio of more intense TC in the WPAC, the intensity distributions were not significantly different in either region when testing the intensity distributions.

In the SHEMI, the number of total TCs has decreased significantly, but the number of Category 3–5 and Category 4–5 TCs increased and these trends were significant at the 95% and 99% confidence level respectively (**Table 2**). The overall decrease was driven by decreases in the number of TS and a decrease in Category 1–2 storms (**Table A1e**) significant at the 99% confidence level (not shown). The 2010–2019 decade showed decreases overall and in the number of hurricanes from the previous decade (2000–2009), and this decade was less active than the last decades of the twentieth century (**Table A1e**). The most recent decrease continued the overall decrease found in Lupo [1]. As for the WPAC however, the percentage of major hurricanes was higher (55%) for the early twenty-first century compared to the late twentieth century (43%).

The EPAC Region showed very little trend throughout the period (**Table A1b, Figure 2e** and **f**) in any category, including no statistically significant trend in the major hurricane categories (**Table 2**). This result is similar to that of Lupo [1]. However, it was clear that the 2010–2019 period in the EPAC was more active than the previous two decades suggesting interdecadal variability. This will be studied below. Additionally, testing the distribution of TC intensities for this region shows

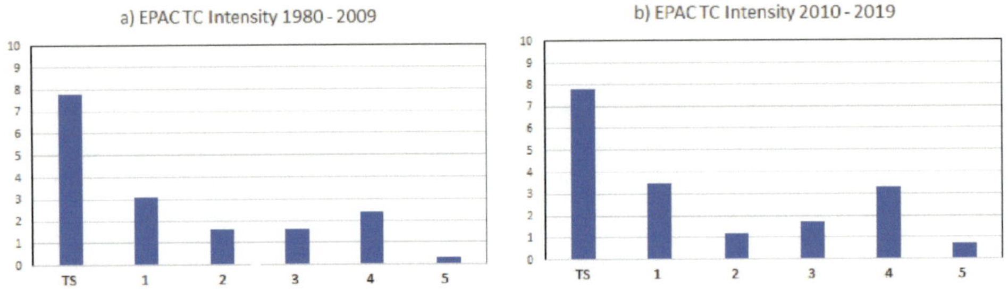

Figure 3.
The TC intensity distributions in the EPAC region for (a) 1980–2009 and (b) 2010–2019. The abscissa is TC intensity and the ordinate is annual occurrence.

that the distribution of TC from 2010 to 2019 was similar to that for the period 1980–2009 at the 95% confidence level when using the χ^2 test (**Figure 3**). This is the only region in which the distributions were similar to an acceptable degree of confidence.

3.2 Interannual and interdecadal variability

In this section, the 2010–2019 results are partitioned by ENSO phase in order to compare these results to those of Lupo [1] and Lupo and Johnston [16]. As shown in Lupo et al. [35], this most recent decade was still classified as a negative PDO. Thus, to examine interannual variability, a comparison was made to the previous decade and interdecadal variability was examined by comparing to the decades of the 1980s and 1990s (**Table A2**). These decades were primarily positive PDO years (1977–1998). This study also provides an opportunity in some ocean basins to compare to the previous negative PDO epoch in order to determine whether the current negative PDO epoch is comparable or if there are differences that may be due to enhanced satellite coverage or if these differences could be physical. The results here were also compared by sub-basin within each global region.

An examination of the Atlantic Region activity (**Table A2a**) demonstrates that there were more TCs observed during LN and NEU years during the latest decade, and this activity was consistent with that of the previous decade. A comparison to the activity during the 1980s and 1990s demonstrated that while there were more TCs overall (significant at the 95% confidence interval when testing the means), the ENSO variability was similar. During each decade, EL years were 30% (or greater) less active than during other years. Thus, there was no significant difference between ENSO variability across the positive phase of the PDO and the current negative phase in spite of a more active negative PDO phase when testing the means in **Table A2a**. Previous studies (e.g., [1]) showed similar results, with the exception that the negative PDO phase showed weaker ENSO-related variability. Additionally, the ENSO variability with respect to TC intensity distributions was similar in that the comparison of the EN years to all years in each phase of the PDO (**Figure 4**)

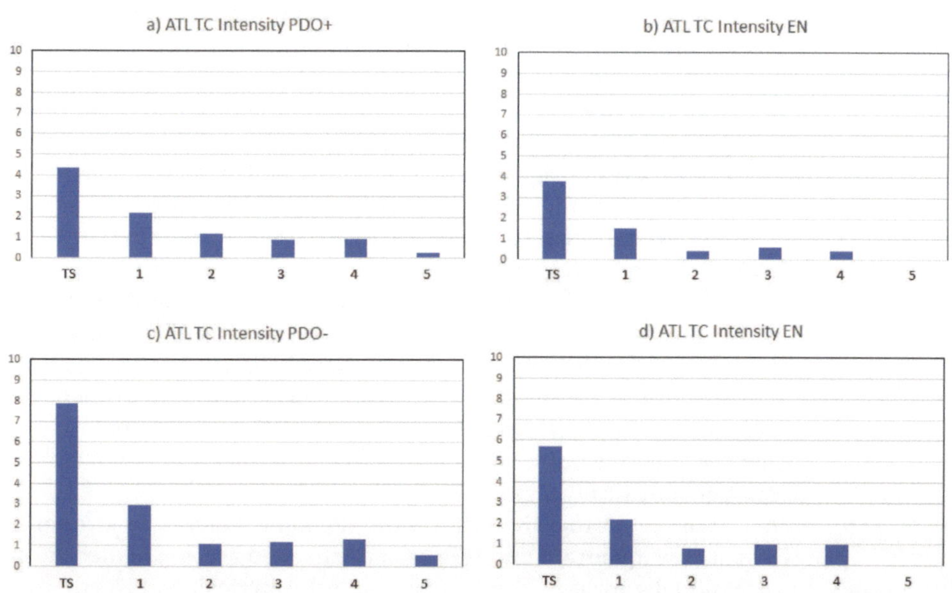

Figure 4.
The TC intensity distributions for (a) all PDO+ TC, (b) all PDO+ EN TC, (c) all PDO − TC, and (d) all PDO − EN TC in the ATL region.

and these were similar at the 90% confidence level. The LN year distributions were different from either the EN years or those overall, but not at standard levels of significance.

A comparison of **Table A2a** to the results of Lupo and Johnston [16] demonstrated that both the current (since 1999) and the previous (1947–1976) negative PDO epoch were more active than the positive PDO epoch (1977–1998). This result is similar to that of Klotzbach and Gray [2], who also show the mid-twentieth century and early twenty-first century were more active times for TC occurrence in the Atlantic Region compared to the latter twentieth century. This also supports the contention of overlap between multi-decadal epochs of the PDO and Atlantic Region teleconnections described in Section 2.3. However, in Lupo and Johnston [16], the number of TCs reaching hurricane strength did not vary at all with respect to ENSO from 1947 to 1976. Their study did not include tropical storms. Thus, it would be difficult to state with certainty that the difference between the results above and the Lupo [1] study are real as they may be a result of not counting TS in the earlier study or the lack of satellite observations. The non-count of TS is supported since if TSs are not included in the current negative PDO period, the ENSO variability in this phase is much weaker.

An examination of the regional occurrence of TC within the Atlantic over the latest decade (**Table A2a**) demonstrates that the western Atlantic is the most active sub-basin and that the ENSO variability within this region is minor. The Gulf and Caribbean sub-basin TC activity was also unchanged as EL years are much less active in these two areas. These results agree with the previous studies from this group and others (see [3]). The only substantial difference between the results presented here and the previous results was that the eastern Atlantic was significantly more active (at the 99% confidence level) even when considering the small sample size. This may be due to increased SSTs over this part of the Atlantic during the last decade (e.g., [40]). A comparison of the length of the TC season (not shown here) to previous results [1] would demonstrate that the Atlantic Region TC season may be beginning about 2 weeks earlier than June 1 as TCs were observed in May for 5 of the 10 years during this decade.

In the East Pacific Region (**Table A2b**), the most recent decade shows ENSO variability that is opposite of the Atlantic Region, in that there are more TCs during EL years than during LN years due to the warmer sea surface temperatures there. This is similar to Collins [24] or Lupo [1]. There was also little difference in TC numbers across the positive and current negative PDO epoch, and their intensity distributions were similar, a result significant at the 95% confidence level (as in **Figure 3**). When taken together, the first two decades of the 40-year period show ENSO variability similar to the latter two decades in that there were about 30–35% fewer TC in LN years. When testing the means, this result was statistically significant at the 95% confidence level. Testing the TC intensity distributions demonstrates that LN years were similar to the overall distributions at the 95% confidence level during both phases of the PDO in a manner similar to **Figure 3**. Even the EN year TC intensity distributions are similar to the overall intensity distribution in the positive PDO phase at the 90% confidence level. Only the EN years TC intensity distribution during the negative phase of the PDO was different from the overall distribution, but not at statistically significant levels.

In **Table A2b**, it is apparent that the East Pacific Region is dominated strongly by activity in the southeast quadrant and this has not changed across any decade or PDO epoch. TC occasionally form further up the Central American and North American coast in the northeast quadrant, but only during LN and NEU years, while TC formed rarely in the northwest quadrant. The southwest quadrant TC activity did account for about 16% of the East Pacific Region activity and the ENSO variability in this quadrant was similar to the previous decades and also similar to that of the southeast quadrant (e.g., Camargo et al. [3] and references therein). The only difference is that the most current decade showed stronger ENSO variability, but this was not statistically significant. Finally, there was no appreciable change in the length of the East Pacific TC season.

As shown above, there has been a decrease in West Pacific hurricanes. **Table A2c** confirms that the TC activity of the most recent decades is less than that of the previous three decades, which can be assumed to be real since satellite coverage has been comprehensive since 1980. However, it is difficult to attribute this decrease to interdecadal variability when comparing to Lupo [1] since the TC activity from the 1940s through the 1970s occurred during an era with less satellite coverage. This same study concluded that there was no significant West Pacific Region interdecadal or interannual variability. Overall, LN years were 20% less active than EN years from 1980 to 2019. For this period and region, this is significant at the 95% confidence level. Thus, there is a strong correlation between the interannual variability within this region and the East Pacific Region (e.g., [3]). An examination of the TC intensity distributions (**Figure 5**) shows that the distribution of negative and positive PDO is similar at the 90% confidence level using the χ^2 test. This is also true for the EN year TC distributions in relation to the intensity distributions for the positive or negative phase of the PDO (**Figure 5**).

Table A2c also demonstrates that the occurrence of TC by quadrant in the West Pacific over the most recent decade was similar to that found in the earlier decades and Lupo [1]. In short, the southwest quadrant is the most active and shows only marginal (insignificant) interannual and interdecadal variability. The southeast quadrant is associated with 30% less TC activity than the southwest, but very strong (statistically significant at the 95% confidence level) ENSO variability. There were two to four times more TCs in the southeast quadrant during EN years, a result that agrees with many studies (e.g., [1, 3]). The recent decreases noted above for the West Pacific Region overall was distributed among the four sub-basins, though as a percentage, the decrease was largest (approximately 35% less) for the northeast quadrant. Additionally, the active southeast quadrant in the West Pacific during EN

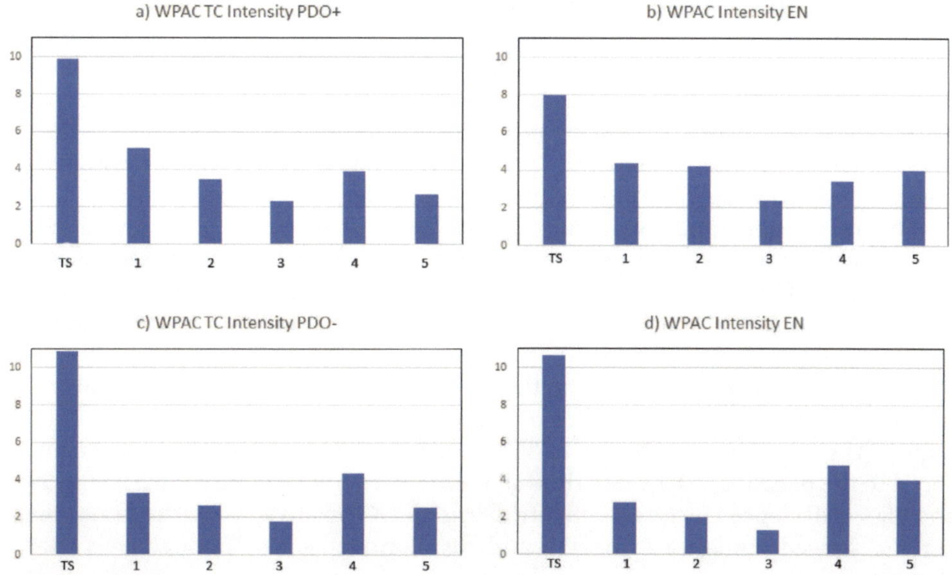

Figure 5.
*As in **Figure 4**, except for the West Pacific region.*

years combined with the active southwest quadrant for the East Pacific supports the conclusions of Camargo et al. [3], Lupo [1] and others in that during El Nino years, the Pacific is active across the basin for EN years, while during LN years activity was centered closer to their respective quadrants for both regions. Like the East Pacific, there was no significant change noted in the length of the TC season here (not shown).

In the North Indian Ocean Basin, there was an increase in the number of TC occurrences as shown above, and **Table A2d** suggests that this was driven primarily by increases in the western Indian Ocean Region including the Arabian Sea. This includes the number of major storms. Since the regional classification for the intensity of these storms began in 1977, there is no need to compare this region or the Southern Hemisphere results (this region began reporting intensity in 1980) to earlier results. **Table A2d** also demonstrated that EN years were slightly more active than other years, and this is opposite that of the previous three decades. Thus, there are no conclusions that can be drawn about ENSO variability, nor about the interdecadal variability. However, Ng and Chan [27] showed that there was strong variability on the 5-year timescale in this region linked to the Indian Ocean Dipole (IOD).

An examination of TC intensity distributions (not shown) shows that regardless of how the results are stratified in the North Indian Ocean Region, the distributions are similar to the overall distribution at the 95% confidence level or higher. The only exception was the distribution of TC intensities in LN years during the positive phase of the PDO were different, but not at standard levels of significance. The reason for the lack of variability in TC intensities in this region may be the less frequent occurrence of storms in this region. Finally, there was no change in the TC season here (not shown) and this was identical to the results of Lupo [1] and references therein who showed that this region possessed a double peak in activity (May–June and October–December), which is associated with the annual migration of the Intertropical Convergence Zone.

The decrease in Southern Hemisphere variability shown in **Table A2e** for the most recent decade (2010–2019) continues the trend identified when comparing 2000–2009 or the previous two decades. Like the NIND Region, there are too few years to attribute these decreases to interdecadal variability as of yet. When examining the sub-basins, the decreases over the past two decades were primarily the result of fewer TC in the East Indian Ocean and to a lesser extent over the southwest Pacific. Additionally, during this decade, the TC season extends from October to April, which is similar to the previous decades [1]. Lastly, an examination of the distribution of TC intensities for the positive versus negative PDO demonstrated the distributions were different, but not at standard levels of significance.

The overall interannual variability during the most recent decade showed more TC during EL years, which was counter to the results of the previous three decades (**Table A2e**). This variability, however, was not significant at acceptable levels of confidence. Lupo et al. (2011) found weak ENSO-related variability, which was marginally significant with more TC occurring during LN years. Examining the sub-basins exposed an error in assigning the ENSO year in Lupo [1] (see their Table 17) for these values only. The overall results were consistent between this study and the Lupo's [1] study. The distribution of TC intensities during EN and LN years compared to the negative PDO years showed these distributions were similar at the 99% and 95% confidence level, respectively. During PDO positive years, the same comparison showed similarity at the 95% and 90% confidence level, respectively (not shown).

A discussion of the SHEMI sub-basin results (**Table A2e**) demonstrates that TC numbers in the West Indian Ocean Basin demonstrate the most current decade was slightly more active than the previous 30 years. EL years were more active over all decades than LN years observing nearly double the TC activity. This is then opposite to what was reported by Lupo [1] as that study reported more TC in LN years. The East Indian Ocean Basin saw the largest decreases as in the previous 30 years, as 11.3 TC events per year were reported. Here, the results show that for the latest decade only 7.7 TCs per year were observed, representing a decrease of about 33%. This was nearly equal to the total decrease in SEMI TC overall. More importantly, in the previous 30 years, LN season TC outnumbered EN season TC by more than two to one. For the latest decade, LN years experienced only 20% more TCs per season. This preference for LN years as in the West Indian Ocean Basin was of the opposite sense reported in Lupo [1]. However, the results presented here now agree with results for the East Indian and West Pacific numbers reported for these regions in earlier studies (e.g., [3], and references therein).

Only in the southwest Pacific were the observed TCs and their interannual variability in the current decade consistent with those of the previous 30 years, showing a slight preference for LN years. Thus, the coding error of Lupo [1] did not have a major impact on the results reported for this sub-basin only. The southeast Pacific was the least active TC region of the SHEMI outside of the South Atlantic, and the occurrences of TC in the latest decade were consistent with the previous three. The latest decade showed only a slight preference for TC occurrences in EL years, and this was consistent with the three previous decades except that the previous decades saw stronger disparities between annual TC occurrences in EL years versus LA years. The ENSO variability in this sub-basin was opposite to what was reported in Lupo [1].

Globally, there were 79.5, 90, and 92 TCs that occurred during LN, NEU, and EN years, respectively, during the last decade. This compares to 82, 91.3, and 85.7 TCs occurring during these years over the previous decade, respectively. The comparative numbers for the 1980–1999 period revealed there were 83.3, 88.8, and 85.6 TCs that occurred per LN, NEU, and EL year, respectively. Thus, the most recent decade demonstrates slightly different ENSO variability from that of the previous three decades, but this difference is not statistically significant. Over the entire 40-year period, these TC occurrence numbers were 81.8, 89.7, and 87.4, during LN, NEU, and EN years respectively.

4. Summary, discussion, and conclusions

In this chapter, the global tropical cyclone activity for 2010–2019 was examined and compared firstly to the TC activity of the previous decade (2000–2009) and then to those occurring from 1980 to 1999. By doing so, we compared the results here to the previous results reported in works such as Camargo et al. [3] or Lupo [1]. The data sources used here were the same as those used in that study. The definitions for the TC season, basins, sub-basins, and internannual and internannual variability were identical to those used in Lupo and Johnston [16] and Lupo [1]. The statistical tests used here can be found in standard statistics text books.

Global TC activity in general during the latest decade was very similar to that of the previous decade and within most sub-basins there were broad similarities as well. However, this study found some key differences from Lupo [1]. The following results are new here.

These are:

- Globally, there were no statistically significant increases or decreases in overall global TC activity although the trend in the number of storms has shown increases. The number of intense storms (Category 3–5) showed a statistically significant increase over the 40-year period similar to IPCC [4], Kossin et al. [11], and others. The number of TSs also increased, but this was not statistically significant. These increases in these TCs were found in most global basins. Only the number of Category 1 and 2 storms decreased, especially since 2000.

- In the ATL Region, the number of TCs during 2010–2019 was similar to 2000–2009. The overall 40-year trend was upward in the total number of TCs, hurricanes only, and intense hurricanes. These were all statistically significant. The interannual variability over the latest four decades was similar in that there were more TCs during LN years (about 30% more). Additionally, the ATL TC season during the 2010–2019 period started about 2 weeks earlier than the previous decades, while the eastern Atlantic observed an increase in TC activity.

- While the intensity distributions were different when comparing negative and positive phases of the PDO, this result was not statistically significant. Also, the distributions of LN and EN TC intensities were compared to the total sample within each phase of the PDO, and the EN intensity distributions were similar at the 90% confidence level.

- In the EPAC, few differences in the climatological character of TC were noted when compared to Collins [24, 25] or Lupo [1]. When comparing the TC intensity, distributions for each phase of the PDO or with respect to ENSO showed that these distributions were similar at standard levels of significance except when comparing the distribution of EN year TC intensities to the distribution of positive PDO TC.

- Other studies showed significant increases in the number of intense TCs within the WPAC. Such an increase was not found here, but significant decreases in the number of Category 1 and 2 storms resulted in an increase in the proportion of WPAC TCs classified as intense. The decrease in the number of TC basin-wide was distributed approximately evenly across each quadrant. In this region, the TC intensity distributions were similar for each phase of the PDO at the 90% confidence level. This same result was found when comparing EN year TC intensities to the total distribution in each PDO phase.

- Within the IND Region, there were significant increases in TC for the latest decade and over the entire 40-year period for total TC occurrence, Category 1 and 2 storms, and intense TCs and all these trends were statistically significant. These increases were especially evident within the western Indian Ocean Basin and Arabian Sea. All TC intensity distributions tested for interannual and interdecadal variability were similar to each other at standard levels of significance.

- In the SHEMI, the 40-year trends showed significant decreases in TC frequency overall including the number of TSs and Category 1 and 2 hurricanes. But there was a significant increase in the number of intense storms. The number of TCs observed over the latest decade was the lowest in the 40-year period and proportion of TCs reaching Category 3 or higher increased. In this region, the positive and negative PDO TC intensity distributions were different, but not at standard levels of significance. The EN and LN year TC intensity distributions in each phase of the PDO were similar to the total sample for that PDO phase.

- There was no significant SHEMI interannual variability overall, but the latest decade showed more TCs in EN years as compared to LN years. This was different from the previous 30 years. A coding error found in the Lupo [1] results showed that the variability associated with ENSO was opposite that reported in Lupo [1] for three of the four sub-basins.

Acknowledgements

The authors are grateful to the anonymous reviewers for their comments on this chapter. Their comments made this work stronger.

Appendix

See Appendix **Tables A1** and **A2**.

Category	1980–1989	1990–1999	2000–2009	2010–2019	1980–2019
a. Atlantic					
TS	4.1	4.6	7.7	8.1	6.1
Cat. 1,2	3.5	3.9	3.7	4.4	3.9
Cat. 3–5	1.7	2.5	3.5	2.7	2.7
Cat. 4,5	1.0	1.4	2.1	1.7	1.6
Tot Hur	5.2	6.4	7.2	7.1	6.5
TS + Hur	9.3	11.0	14.9	15.2	12.6
b. East Pacific					
TS	8.6	5.6	9.1	7.8	7.8
Cat. 1,2	5.4	4.5	4.4	4.7	4.7
Cat. 3–5	4.6	5.5	2.8	5.7	4.7
Cat. 4,5	2.3	3.9	1.8	4.0	3.0
Tot Hur	10.0	10.0	7.2	10.4	9.4
TS + Hur	18.6	15.6	16.3	18.2	17.2
c. West Pacific					
TS	9.7	10.3	10.4	11.3	10.4
Cat. 1,2	8.1	9.1	6.5	5.1	7.2
Cat. 3–5	8.4	9.0	9.5	8.3	8.8
Cat. 4,5	5.4	7.3	8.0	6.3	6.8
Tot Hur	16.5	18.1	16.0	13.4	16.0
TS + Hur	26.2	28.4	26.4	24.7	26.4
d. North Indian Ocean					
TS	3.7	2.9	4.0	3.2	3.4
Cat. 1,2	0.5	1.3	0.6	1.1	0.9
Cat. 3–5	0.2	1.0	0.5	1.2	0.7
Cat. 4,5	0.1	0.7	0.4	0.8	0.5

Category	1980–1989	1990–1999	2000–2009	2010–2019	1980–2019
Tot Hur	0.7	2.3	1.1	2.3	1.6
TS + Hur	4.4	5.2	5.1	5.5	5.0
e. Southern Hemisphere					
TS	14.6	12.8	13.0	12.0	13.1
Cat. 1,2	8.0	7.8	5.6	5.3	6.7
Cat. 3–5	5.0	7.3	7.1	6.3	6.4
Cat. 4,5	1.6	4.8	4.5	4.8	3.9
Tot Hur	13.0	15.1	12.8	11.6	13.1
TS + Hur	27.6	27.9	25.8	23.6	26.2
f. Global					
TS	40.7	37.3	44.4	43.7	41.3
Cat. 1,2	25.5	26.6	21.0	20.9	23.4
Cat. 3–5	19.8	25.3	23.2	24.2	23.2
Cat. 4,5	10.4	18.1	16.7	17.6	15.6
Tot Hur	45.4	51.9	44.3	44.8	46.6
TS + Hur	86.1	88.2	88.7	88.5	87.9

Table A1.
The decadal mean number of tropical storm (TS), category 1–2, category 3–5, category 4–5, total hurricanes, and total TC for each decade from the 1980s to the 2010s and for the entire period within each global ocean basin and over the entire globe.

a. Atlantic	**All**	**CRBN**	**GULF**	**WATL**	**EATL**
1980–1999/2000–2009					
LN (3/1)	12.7/15.0	2.3/2.0	3.0/4.0	3.7/7.0	3.7/2.0
NEU (12/6)	10.1/17.3	1.0/3.3	1.9/3.5	5.8/7.0	2.2/3.7
EN (5/3)	7.0/10.0	0.4/1.3	1.4/1.7	4.0/6.3	1.2/1.3
Total	10.2/14.9	1.1/2.6	1.9/3.0	5.1/6.8	2.2/2.8
2010–2019					
LN (2)	18.0	4.0	2.0	6.5	5.5
NEU (5)	16.2	1.6	3.8	6.5	4.4
EN (3)	11.3	1.3	0.7	5.3	4.0
Total	15.2	2.0	2.5	6.2	4.5
b. East Pacific	**All**	**NW**	**NE**	**SW**	**SE**
1980–1999/2000–2009					
LN (3/1)	12.3/14.0	0.0/0.0	1.7/1.0	1.7/3.0	9.0/10.0
NEU (12/6)	17.9/15.7	0.1/0.2	1.1/0.8	2.8/2.2	13.8/12.5
EN (5/3)	18.4/18.3	0.0/0.0	0.4/0.3	3.2/4.0	14.6/14.2
Total	17.1/16.3	0.1/0.1	1.0/0.7	2.7/2.8	13.3/12.7
2010–2019					
LN (2)	13.0	0.5	0.5	0.5	11.5

NEU (5)	17.2	0.0	0.4	2.0	14.8
EN (3)	23.3	0.0	0.0	5.7	17.7
Total	18.2	0.1	0.3	2.8	15.0
c. West Pacific	**All**	**NW**	**NE**	**SW**	**SE**
1980–1999/2000–2009					
LN (3/1)	22.7/25.0	3.7/4.0	3.3/6.0	13.0/11.0	2.7/4.0
NEU (12/6)	28.4/26.7	2.6/3.0	3.7/3.3	13.0/13.2	9.2/7.2
EN (5/3)	27.4/26.3	2.2/3.3	1.6/3.0	10.8/9.3	12.8/10.7
Total	27.3/26.4	2.7/3.2	3.1/3.6	12.5/11.8	9.1/7.9
2010–2019					
LN (2)	20.1	4.0	1.0	11.0	4.0
NEU (5)	26.2	2.6	3.4	12.6	7.6
EN (3)	25.0	2.7	1.0	10.3	11.0
Total	24.7	2.9	2.2	11.6	8.0
d. North Indian	**All**	**West**	**East**		
1980–1999/2000–2009					
LN (3/1)	6.0/6.0	2.3/2.0	3.7/4.0		
NEU (12/6)	4.5/5.0	1.2/1.5	3.3/3.5		
EN (5/3)	4.8/5.0	0.8/1.3	4.0/3.7		
Total	4.8/5.1	1.3/1.5	3.6/3.6		
2010–2019					
LN (2)	4.5	1.5	3.0		
NEU (5)	5.8	2.6	3.2		
EN (3)	5.7	3.0	2.7		
Total	5.5	2.5	3.0		
e. Southern Hemisphere	**All**	**W IND**	**E IND**	**SW PAC**	**SE PAC**
1980–1999/2000–2009					
LN (2/2)	30.5/28.0	4.0/5.5	15.0/14.5	8.0/5.5	3.5/2.5
NEU (13/6)	27.8/25.0	6.1/5.7	12.5/11.8	7.2/4.7	1.9/2.8
EN (5/2)	26.6/26.0	6.6/9.5	6.8/6.5	6.6/6.5	6.6/3.5
Total	27.8/25.8	6.0/6.4	11.4/11.3	7.1/5.2	3.3/2.9
2010–2019					
LN (2)	22.5	4.0	8.5	7.5	2.5
NEU (4)	22.8	6.5	8.3	4.8	3.3
EN (4)	25.2	8.5	6.8	6.8	3.0
Total	23.6	6.8	7.7	6.1	3.0

Table A2.
The mean annual TC occurrence stratified by ENSO phase and sub-basin for the (a) ATL, (b) EPAC, (c) WPAC, (d) NIND, and (e) SHEMI.

Author details

Anthony R. Lupo*, Brendan Heaven, Jack Matzen and Jordan Rabinowitz
School of Natural Resources, University of Missouri, Columbia, MO, USA

*Address all correspondence to: lupoa@missouri.edu

References

[1] Lupo AR. The interannual and interdecadal variability in hurricane activity, Chapter 1. In: Lupo AR, editor. Recent Hurricane Research: Climate, Dynamics, and Societal Impacts. Rijeka, Croatia: IntechOpen; 2011. pp. 38, 616

[2] Klotzbach PJ, Gray WM. Multidecadal variability in North Atlantic tropical cyclone activity. Journal of Climate. 2008;**21**:3929-3935

[3] Camargo SJ, Sobel AH, Barnston AG, Klotzbach PJ. The influence of natural climate variability on tropical cyclones, and seasonal forecasts of tropical cyclone activity. Global Perspectives on Tropical Cyclones, World Scientific Series on Asian Pacific Weather and Climate. 2010;**4**:325-360

[4] Intergovernmental Panel on Climate Change (IPCC). Climate Change 2013: The Physical Scientific Basis. 2013. Available from: http://www.ipcc.ch

[5] Walsh KJE, McBride JL, Klotzbach PJ, Balachandran S, Camargo SJ, Holland G, et al. Tropical cyclones and climate change. Wiley Interdisciplinary Reviews: Climate Change. 2015;7:65-89. DOI: 10.1002/wcc.371

[6] Zhao J, Zhan R, Wang Y. Global warming hiatus contributed to the increased occurrence of intense tropical cyclones in the coastal regions along East Asia. Nature, Scientific Reports. 2018;**8**:6023. DOI: 10.1038/s41598-018-24402-2

[7] Mei W, Xie SP. Intensifcation of landfalling typhoons over the Northwest Pacific since the late 1970s. Nature Geoscience. 2016;**9**:753-757

[8] Elsner JB, Kossin JP, Jagger TH. The increasing intensity of the strongest tropical cyclones. Nature. 2008;**455**: 92-95

[9] Holland G, Bruyère CL. Recent intense hurricane response to global climate change. Climate Dynamics. 2014;**42**:617-627

[10] Liu M, Vecchi GA, Smith JA, Knutson TR. Causes of large projected increases in hurricane precipitation rates with global warming. npj Climate and Atmospheric Science. 2019;**2**:38. DOI: 10.1038/s41612-019-0095-3

[11] Kossin JP, Knapp KR, Olander TL, Velden CS. Global increase in major tropical cyclone exceedance probability over the past four decades. Proceedings of the National Academy of Sciences of the United States of America. 2020. DOI: 10.1073/pnas.1920849117

[12] Wang C, Wang X, Weisberg RH, Black ML. Variability of tropical cyclone rapid intensification in the North Atlantic and its relationship with climate variations. Climate Dynamics. 2017;**49**:3627-3645. DOI: 10.1007/s00382-017-3537-9

[13] Emanuel KA. The dependence of hurricane intensity on climate. Nature. 1987;**326**:483-485

[14] Zuki MZ, Lupo AR. The interannual variability of tropical cyclone activity in the southern South China Sea. Journal of Geophysical Research. 2008;**113**:D06106. DOI: 10.1029/2007JD009218-14

[15] Henderson-Sellers A et al. Tropical cyclones and global climate change: A post-IPCC assessment. Bulletin of the American Meteorological Society. 1998; **79**:19-38

[16] Lupo AR, Johnston GJ. The variability in Atlantic Ocean Basin hurricane occurrence and intensity as related to ENSO and the North Pacific Oscillation. National Weather Digest. 2000;**24**:1-13

[17] Roberts MJ, Vidale PL, Mizielinski MS, Demory ME, Schiemann R, Strachan J, et al. Tropical cyclones in the UPSCALE ensemble of high-resolution global climate models. Journal of Climate. 2015;**28**:574-596

[18] Strachan J, Vidale PL, Hodges K, Roberts M, Demory ME. Investigating global tropical cyclone activity with a hierarchy of AGCMs: The role of model resolution. Journal of Climate. 2013;**26**: 133-152. DOI: 10.1175/JCLI-D-12-17 00012.1

[19] Walsh KJE, Camargo SJ, Vecchi GA, Daloz AS, Elsner J, Emanuel K, et al. Hurricanes and climate: The U.S. CLIVAR working group on hurricanes. Bulletin of the American Meteorological Society. 2015;**96**(6):997-1017. DOI: 10.1175/BAMS-D-13-00242.1

[20] Kossin JP, Olander TL, Knapp KR. Trend analysis with a new global record of tropical cyclone intensity. Journal of Climate. 2013;**26**:9960-9976

[21] Dvorak VF. A technique for the analysis and forecasting of tropical cyclone intensities from satellite pictures. NOAA Technical Memorandum NESS 45. 1973. p. 19

[22] Dvorak VF. Tropical cyclone intensity analysis using satellite data. NOAA Tech. Rep. NESDIS 11. Washington, DC: NOAA/NESDIS; 1984. p. 47

[23] Knaff JA, Brown JP, Courtney J, Gallina GM, Beven JL. An evaluation of Dvorak technique–based tropical cyclone intensity estimates. Weather and Forecasting. 2010;**25**:1362-1379

[24] Collins JM. The relationship of ENSO and relative humidity to interannual variations of hurricane frequency in the North-East Pacific Ocean. Papers of the Applied Geography Conference. 2007;**30**:324-333

[25] Collins JM. Contrast high north-East Pacific tropical cyclone activity with low North Atlantic activity. Southeastern Geographer. 2010;**50**(1):83-98

[26] Knapp KR, Kruk MC, Levinson DH, Diamond HJ, Neumann CJ. The international best track archive for climate stewardship (IBTrACS). Bulletin of the American Meteorological Society. 2010;**91**:363-376

[27] Ng KWE, Chan JCL. Interannual variations of tropical cyclone activity over the northern Indian Ocean. International Journal of Climatology. 2012;**32**:819-830

[28] Landsea CW. A climatology of intense (or major) Atlantic hurricanes. Monthly Weather Review. 1993;**121**: 1703-1713

[29] Lupo AR, Latham TK, Magill T, Clark JV, Melick CJ, Market PS. The interannual variability of hurricane activity in the Atlantic and East Pacific regions. National Weather Digest. 2008; **32**(2):119-135

[30] Vecchi GA, Knutson TR. Estimating annual numbers of Atlantic hurricanes missing from the HURDAT database (1878–1965) using ship track density. Journal of Climate. 2011;**24**:1736-1746

[31] Simpson RH. The hurricane disaster potential scale. Weatherwise. 1974; **27**(169):186

[32] Neter J, Wasserman W, Whitmore GA. Applied Statistics. 3rd ed. Boston: Allyn and Bacon; 1988. p. 1006

[33] Wilks DS. Statistical Methods in the Atmospheric Sciences. International Geophysics Series Number 91. 2nd ed. Cambridge, MA, USA: Academic Press; 2006. p. 627

[34] Lupo AR, Oglesby RJ, Mokhov II. Climatological features of blocking

anticyclones: A study of northern hemisphere CCM1 model blocking events in present-day and double CO_2 atmospheres. Climate Dynamics. 1997; **13**:181-195

[35] Lupo AR, Jensen AD, Mokhov II, Timazhev AV, Eichler T, Efe B. Changes in global blocking character in recent decades. Atmosphere. 2019;**10**:92

[36] Pielke RA Jr, Landsea CN. La Nina, EI Nino, and Atlantic hurricane damages in the United States. Bulletin of the American Meteorological Society. 1999;**80**:2027-2034

[37] Hanley DE, Bourassa MA, O'Brien JJ, Smith SR, Spade ER. A quantitative evaluation of ENSO indices. Journal of Climate. 2003;**16**: 1249-1258

[38] Mantua NJ, Hare SR, Zhang Y, Wallace JM, Francis RC. A Pacific Interdecadal climate oscillation with impacts on Salmon production. Bulletin of the American Meteorological Society. 1997;**78**:1069-1079

[39] Minobe S. A 50–70 year climatic oscillation over the North Pacific and North America. Geophysical Research Letters. 1997;**24**:683-686

[40] Prigent A, Lübbecke JF, Bayr T, et al. Weakened SST variability in the tropical Atlantic Ocean since 2000. Climate Dynamics. 2020;**54**:2731-2744. DOI: 10.1007/s00382-020-05138-0

Section 2

Tropical Cyclone Dynamics

Chapter 2

Impacts of Tropical Cyclones in the Northern Atlantic on Adverse Phenomena Formation in Northeastern Brazil

Natalia Fedorova, Vladimir Levit and Lucas Carvalho Vieira Cavalcante

Abstract

Tropical cyclone (TC) impacts on adverse phenomena in the tropical region of Northeastern Brazil (NEB) have been analyzed. TC influence on fog and rain formation was not described in the previous papers. The main goal of the chapter is to evaluate the existence of such influence and thus to improve the weather forecasting in this area. TC information from the NHC of the NOAA was used. METAR and SYNOP data were used for the adverse phenomena study. Analysis of the synoptic systems was based on different maps at the pattern levels and on satellite images. These maps were elaborated using reanalysis data from the ECMWF. Thermodynamic analysis was also used. Middle tropospheric cyclonic vortexes (MTCV) in the tropical region of the Southern Atlantic were described recently. Five from 10 MTCVs were associated with tropical cyclones and disturbances in the Northern Atlantic. Circulation patterns between TC and synoptic systems at the NEB are described. These circulations create sinking over the BNE and, as a result, form fog, mist and weak rain in the BNE during TC days. Mechanisms of TC influence on weather formation in the BNE are presented. This information is important for improving weather forecasting methods.

Keywords: tropical cyclones, Northeast Brazil, adverse phenomena

1. Introduction

Fog and mist events are rare and show significant variation of frequency in the tropical region of the Brazilian Northeast (BNE) [1, 2]. Radiosonde data in the central and south regions of the BNE, i.e., Recife, Pernambuco state (8°S, 35°W), and Salvador, Bahia state (13°S, 39°W), registered fog (haze) on average in 13 (13) and 37 (41) days per year, respectively [3]. On the other hand, in the first study of fog formation in Maceio, Alagoas State (9°S, 39°W) [4], only two fog events (moderate and weak) during 1996 (both in winter) were found. Fog does not exist in the semiarid region [3]. In the northern region of the BNE, i.e., Belem, Para State (1°S, 48°W), and Quixeramobim, Ceara State (5°S, 39°W), fog (haze) was registered in

24 (1) and 4 (28) events/year, respectively. More frequently, fog was observed in Barra do Corda, Maranhao State (5°S, 45°W), in 48 days/year, but haze was very rare, only 4 days/year. Also, fog duration was very short, 1–2 h on average in the coastal region of the BNE [5].

Moreover, the physical mechanism of fog formation on the northern coast was atypical (typical radiation or advection fog does not occur typically) [6]. Summarizing the results of this chapter, it was possible to make the following conclusion regarding the processes of fog formation in the tropical region of the BNE. A weak confluence at the low-level trough (wave disturbances in trade winds—WDTW) contributes to weak pressure anomaly and creates the conditions for moisture convergence as well. A change in wind direction and air current from the river region contributes additional humidity for fog formation. Also, precipitation occurred before the fog events, contributing to humidification of the air through evaporation of the humid surface and raindrops. A warmer sea surface contributes to more evaporation and, as a consequence, increases the amount of water vapor in the surrounding air at the low levels near the coast. Positive latent heat flux shows a humidity increase and, therefore, moisture accumulation in the coastal region. Negative sensible heat flux results in air cooling, possible water vapor condensation, and, finally, fog formation.

All this information about rare fog events with a short duration and atypical physical mechanisms of their formation demonstrates the numerous problems for fog/haze forecasting.

An intertropical convergence zone (ITCZ), a South Atlantic subtropical high (SASH), trade winds, and an upper tropospheric cyclonic vortex are typical synoptic scale systems associated with different weather conditions in the BNE [7].

For the first time, the influence of wave disturbances in trade winds (WDTW) on rain formation in the BNE was shown in Molion and Bernardo [8]. The relationship between different types of troughs and adverse meteorological phenomena (fog and thunderstorms) in Alagoas State was studied by Rodrigues et al. [9]. It was noted that 87% of the troughs were associated with wave disturbances in trade winds (WDTW) on the northwestern periphery of the subtropical South Atlantic High. Rare stratus cloud events with a duration of 4 days each on the northeast coast of Brazil also were formed in WDTW [10]. Fog formation is usually associated with WDTW in Maceio (Alagoas State) [1, 5, 6, 11].

The influence of the Brazilian Northeast Jet Stream (BNEJS) on weather in the BNE was described recently [12–14]. Particularly, the results of this study show that the period from April to October (a rainy period and the transition to a dry season in the coastal region) was characterized by a rather high number of fast BNEJSs, with high wind speed in the core, a predominant northwesterly direction, and the location of BNEJS between the upper tropospheric trough (UTT) and the SASH.

Upper tropospheric cyclonic vortices (UTCVs) are important synoptic scale systems in this region; the greatest UTCV frequency occurs during the austral summer period, with the maximum frequency in January [15–18]. Intensive cloud development with precipitation was observed on the UTCV periphery and cloudlessness in their center as a result of a downward motion in its cold center and upward motion on the periphery. Rao and Bonatti [18] conjectured that barotropic instability of the regional mean basic winds at the upper levels could be one cause for UTCV formation. It was suggested that these air streams are associated with the subtropical jet stream and contribute to UTCV development. A strong positive shear zone was formed within the South Atlantic trough before the formation of the UTCV [19]. The same paper showed that the barotropic process dominates over the baroclinic one. They showed that barotropic energy conservation is the energy source for the growth of wave motion and wave amplitude and also that barotropic instability of the shear zone can excite the UTCV.

Middle tropospheric cyclonic vortices (MTCVs) located between 700 and 400 hPa in the BNE were described recently [20, 21]. About 232 MTCVs were observed each year over the BNE and adjacent ocean region. MTCVs were predominantly short and lasted less than 12 h. The vortices persisting longer than 30 h were detected more frequently in the summer and rarely in autumn.

Frontal zones are observed regularly in the southern region of the Brazilian Northeast in the Southern Hemisphere (SH) [22, 23]. The influence of a western edge of a frontal cloud band on the weather conditions was found in the central region of the Brazilian Northeast SH and was described by Fedorova et al. [24]. The western edge of a frontal cloud band rarely passes across Alagoas State in the BNE. Only two to five frontal zones per year which directly affected the weather conditions were observed in 2004–2006. Nonetheless, Reeder and Smith [25] quoted different papers (including [26, 27]) where a cold front crossing the equator and penetrating the Northern Hemisphere (NH) tropics was documented. These fronts can initiate severe convection in the subtropic SH; however, they generally tend to suppress convection in the ITCZ.

A SASH was observed in the South Atlantic during the whole year and was located at the low levels in the eastern part of the South Atlantic, on average between 20°W and 30°S [7, 28]. It migrated from 15°W, 27°S in August, to 5°W, 33°S in February [29]. A SASH in summer is frequently split and is normally weaker [7].

Some studies show the influence of synoptic scale systems of the Northern Hemisphere on BNE weather. The influence of tropical cyclone Dany-15 on intensive fog formation in the NEB was descried by Fedorova and Levit [1]. But the information about tropical cyclone (TC) impact on BNE weather over many years has not yet been presented. Thus, the main aim of this chapter is to analyze the impact of all tropical cyclones formed from 2013 to 2015 between 20°N and the equator and which passed through 35–50°W, on fog, mist, and weak rain formation in the BNE.

2. Data and methodology

2.1 Selection and classification of the tropical cyclones

All tropical cyclones (TC) formed from 2013 to 2015 between 20°N and the equator whose centers passed through 35–50°W have been studied, using data from the National Hurricane Center's Tropical Cyclone Reports https://www.nhc.noaa.gov/data/tcr/.

TC has been classified according to their sustained wind speed [30] as follows:

Tropical disturbance (*TD*). In this stage, winds are less than 17 m/s with open circulation (no closed isobars).

Tropical depression (*TDep*). Winds are less than 17 m/s, but there is a closed circulation (closed isobars).

Tropical storm (*TS*). Winds are greater than or equal to 17 m/s but less than 33 m/s with a definite closed circulation. The storm is usually assigned a name.

Hurricane (*H*). Winds greater than or equal to 33 m/s.

Major hurricane (*MH*). Typhoon with 1 min of sustained winds of 65 m/s or greater (*MH* was not observed during the study period).

2.2 Analysis of meteorological phenomenon in the BNE

Fog, mist, and weak rain events in the BNE were analyzed by SYNOP and METAR data. The BNE area and location of all meteorological stations are presented in **Figure 1**.

2.3 Synoptic and thermodynamic analysis

Synoptic analysis was elaborated in the equatorial and tropical regions of both hemispheres between 30°N and 30°S and by latitude between 10°W and 60°W. This region includes the BNE, part of the South Atlantic and the TCs moving over the North Atlantic region.

The tropospheric structure for all TC events has been studied using data from the European Center for Medium-Range Weather Forecasts (ECMWF), with a resolution of 0.25°× 0.25°, 00, 12, 18 UTC (http://apps.ecmwf.int/datasets/data/interim-full-daily/levtype=pl). Streamlines have been elaborated at the low (1000 hPa), middle (500 hPa), and high levels (300 and 200 hPa). Divergence maps were constructed only at the high levels (300 and 200 hPa). Both horizontal maps and vertical sections were used for vertical movement identification. These vertical sections passed through the TC center and the BNE region at the same longitude. Satellite infrared images from the GOES-13 and METEOSAT-10 (http://bancodedados.cptec.inpe.br) were used for synoptic system identification.

Thermodynamic analysis was elaborated for the days with fog, mist, and weak rain in the BNE. Vertical profiles were constructed using the same data from the ECMWF quoted above.

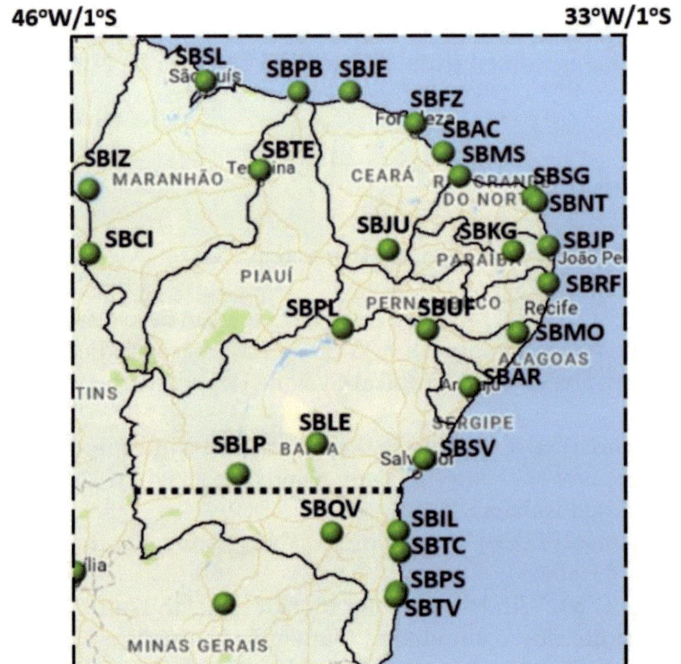

Figure 1.
The location of the BNE, States, and all meteorological stations (green points) with names.

3. Results

3.1 Information about tropical cyclones

Information about tropical cyclones (TC) formed during 2013–2015 between 20°N and the equator and passing through 35–50°W is presented in **Table 1**. Ten tropical cyclones were registered during this time period, and their influences have been analyzed. Five TCs were observed in 2015, three of them in 2013 and only two in 2014.

3.2 Information about fog, mist, and rain formation by METAR and SYNOP data

Information about fog, mist, and weak rain events in the BNE by METAR data when TCs passed through 35–50°W is presented in **Figure 2** and **Table 2**. Two examples of maps with this information are presented in **Figure 2**. Fog, mist. and weak rain events were observed more frequently in the coastal region, e.g., **Figure 2**. Only one fog event was registered far from the coastal region, occurring in the Maranhao State (SBIZ station) (**Figure 2a**). Fog, mist, and weak rain events were observed more frequently in the southern region of the BNE, in the Bahia State. Six events were registered in the southern region of the Bahia (SBQV), not far from the coastal region (300 km approximately). Also, fog was detected in the coastal stations of this region (SBIL and SBPS). Only one event was recorded in the northern coastal region in the Pernambuco State (SBRF). An influence of seven TCs on fog rain formation in the BNE was confirmed (**Table 2**). During three TCs (Edouard, Fred, and Grace), METAR data did not record any fog events; at the same time, weak rain and mist were detected. A *Tropical disturbance and Tropical storm* were the predominant TC stages during its influence on the weather in the BNE.

Name	Data			TC intensity	
	Formation	Dissipation	BNE longitude	Maximal	BNE longitude
2013					
Chantal	06 July	10 July	7–10 July	*TS*	*TS*
Dorian	22 July	03 August	24–27 July	*TS*	*TS*
Erin	15 August	20 August	16–20 August	*TS*	*TS-D*
2014					
Bertha	29 July	06 August	29–31 July	*H*	*TD*
Edouard	10 September	23 September	10–14 September	*H*	*TD-H*
2015					
Danny	17 August	24 August	17–22 August	*H*	*TD-H*
Erika	24 August	28 August	24–25 August	*TS*	*TS*
Fred	30 August	06 September	02–05 September	*H*	*TS-TDep*
Grace	05 September	09 September	06–09 September	*TS*	*TS-D*
Ida	15 September	28 September	17–28 September	*TS*	*TD-D*

TD, tropical disturbance; TDep, tropical depression; TS, tropical storm; H, hurricane; D, dissipation. Source: National Hurricane Center, 2017.

Table 1.
Information about time period and intensity of tropical cyclones (TC).

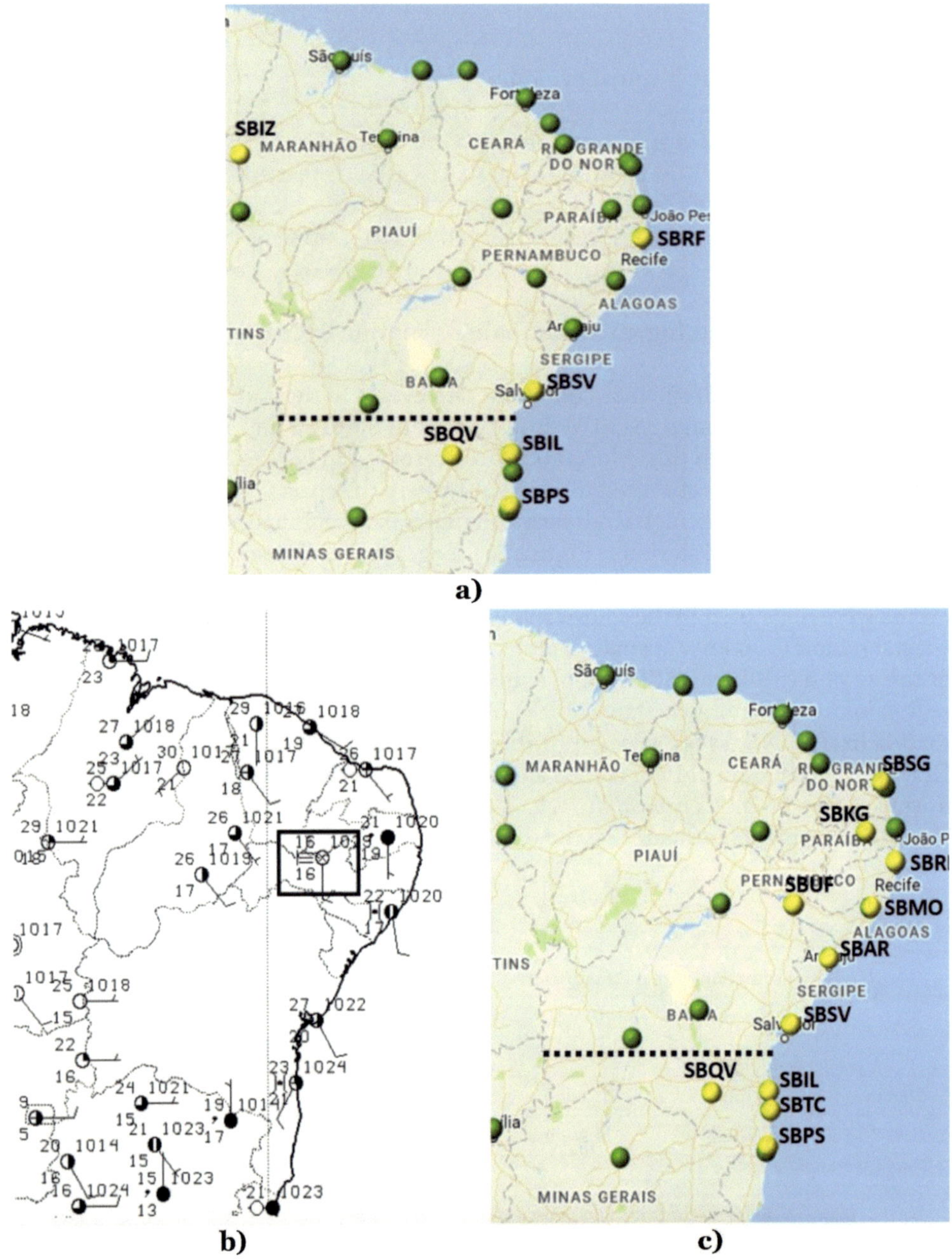

Figure 2.
(a) Location of the meteorological stations detected fog (yellow points) in the BNE by METAR during all TC events. (b) and (c) Location of the meteorological stations detected mist events (yellow points) 29 July 2014 12 UTC in the BNE by METAR (c) and SYNOP (b). Green points mark all meteorological stations.

3.3 Synoptic systems associated with fog events

3.3.1 1000 hPa

A synoptic situation at 1000 hPa was very similar during all fog, mist, and weak rain events (**Table 3**). Trade winds presented cyclonic curvature (*weak trough*) over the ocean close to the coastal area and anticyclonic curvature (*weak ridge*) over the

TC name, *stage*	Data	State	Station	Hours (local)
2013				
Chantal *TS*	08 July	Bahia South	SBQV	08, 09
Dorian *TS*	**22** July	Bahia South	SBQV	10, 11
	25 July	Paraiba	SBKG	06, 09, 10
	25 July	Maranhao	SBIZ	08
	26 July	Paraiba	SBKG	03–08
		Paraiba		*06*
Erin *TS*	**15** August	Bahia North	SBSV	08–10
	18 August	Pernambuco	SBRF	15
		Pernambuco		*12*
2014				
Bertha *TD*	*29 July*	*Pernambuco*		*12*
	31 July	Bahia South	SBQV	10, 11
	01 August	Bahia South	SBQV	09
Edouard	—	—	—	
2015				
Danny *TD*	*17 August*	*Pernambuco*		*12*
	20 August	Bahia South	SBQV	09, 10
	21 August	Paraiba	SBKG	04, 05
	21 August	Bahia South	SBQV	10, 11
Erika *TS*	24 August	Bahia South	SBIL	09, 10
Fred	—	—	—	—
Grace	—	—	—	—
Ida *TD*	18 September	Bahia South	SBPS	09

Bold data show the information when a TC passes before or after the BNE longitudes. TD, tropical disturbance; TS, tropical storm. Source: DECEA, 2017.

Table 2.
Information (TC name, state, station, and hour) about fog events in the BNE by METAR data and SYNOP data (inclined) during the time period with TC passing through 35–50°W.

northern region of the BNE (**Figure 3**). This weak trough was located close to the frontal extremity over the south of the Bahia State in five events. It is important to note that a frontal extremity does not have a direct influence in the fog region but has action on trough formation.

3.3.2 500 hPa

Anticyclonic circulation over the BNE was the predominant process at the middle levels in days with fog, mist, and weak rain events (**Figure 4**, **Table 3**). Also, a convergence zone was observed in the north and middle regions of the BNE in all events. This convergence zone was formed between the currents from a tropical cyclone or tropical depression in the Northern Hemisphere and this anticyclonic circulation in the Southern Hemisphere. The current from the NH was formed as a result of the divergence at the middle levels of the tropical cyclone or tropical depression. For example, **Figure 3** shows airflow from the tropical depression on the west coast of Africa (6°N) converging with the air current of the high (with the center at 6°S, 54°W).

TC name, *stage*	1000 hPa	500 hPa	200/300 hPa
2013			
Chantal *TS*	*TO/RC* (7), *TO/RC/FZ* (8,9)	A (7–9)	*TCO* (6, 7) A (8) *TCO* (9) SJSHS 10–15°
Dorian *TS*	*TO/RC/FZ* (22–26)	A (22–25), MTCVo (23–25)	A (22–23) *TCO* (24, 25) A (26, 27) SJSHS 20°
Erin	*TO/RC/FZ* (16–17)	MTCVc, A (16–18)	A (16) *TCO* (17) A (18) SJSHS 20°
2014			
Bertha *TDis*	*TO/RC/FZ* (29–31–1)	A (29–31–1)	*TCO* (29–31–1), SJSHS 10–15°
Edouard	*TO/RC* (10–13)	A (10), MTCVc (10, 11)	*TCO* (10–11) A (12–13) SJSHS 10–15°
2015			
Danny *TDis*	*TO/RC* (17–19), *TO/RC/FZ* (20–21)	A, MTCVc (17–18), A (19–22)	*TCO* (17–20) SJSHS 10–15° (17–20)
Erika	*TO/RC/FZ* (24–26)	A (24–26)	*TCO* (24–26) SJSHS 10–15° (24–26)
Fred	*TO/RC* (1) *TO/RC/FZ* (2–3)	A (1–4)	*TCO* (1–2) A (3–4) SJSHS 10–15°
Grace	*TW(6–9)*	A (7–9)	*TCO* (7) A (8–9) SJSHS 10–15°
Ida	*TO/RC* (16–18)	A (16–17) MTCVo A (16–17)	A (17–19)

A, anticyclonic circulation; TO (trough over the ocean); RC (ridge over the continent); TW, trade winds nearly straight (no pronounced curvature); FZ (frontal zone), TCO (trough at the middle and high levels over the continent and ocean); SJSHS (subtropical jet stream of the Southern Hemisphere)—location of the northern boundary (°W); MTCVo and MTCVc—middle tropospheric cyclonic vortices over the oceanic and continental regions, respectively. Information inside brackets represents number of days.

Table 3.
Synoptic systems in the BNE at the low (1000 hPa), middle (500 hPa), and high (300 and 200 hPa) levels during tropical cyclones in the northern hemisphere.

Middle tropospheric cyclonic vortices (MTCV) were observed in five studied events (**Table 3**). MTCVs were located over the adjacent ocean region in two of the five events, in the northeastern part and in the coastal region of the continent in one case each. Four MTCVs were typical in terms of the level of its location and duration, were short, and lasted 6–12 h. Only one MTCV lasted 36 h, and its center was located in the south of the BNE, and its circulation affected only southern part of the BNE. The MTCVs created the waves on the air current from the NH but did not change the principal influence of this northern current. Meanwhile one MTCV on 17 Ag 2013, 12UTC, was formed between west flow of the SH and

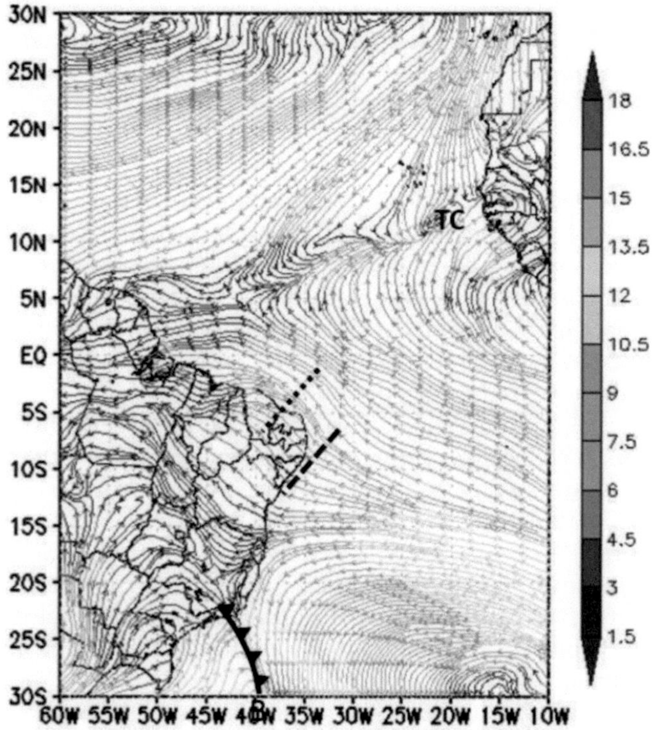

Figure 3.
Streamlines at 1000 hPa on 22 July 2013, 18 UTC. Trough, dashed line; ridge, dotted line; and TC, tropical cyclone.

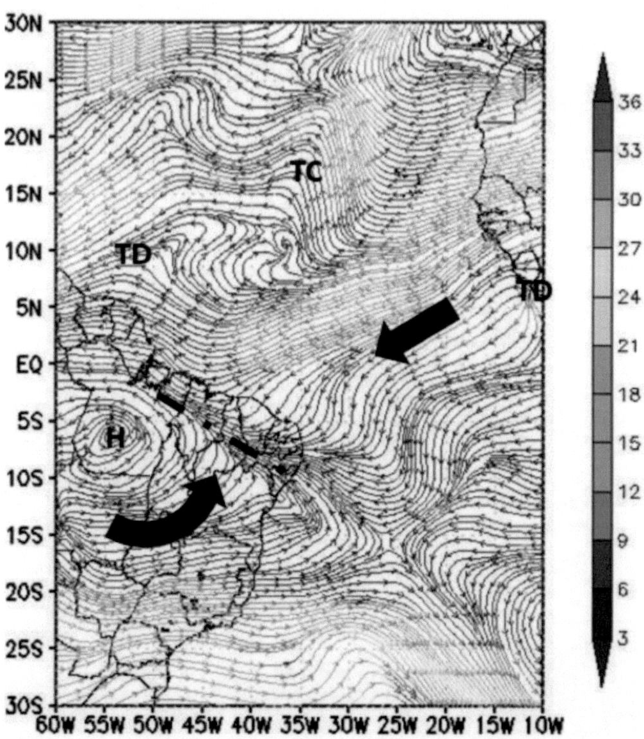

Figure 4.
Streamlines at 500 hPa, 17 August 2013, 00UTC. H, anticyclonic circulation; TD, tropical disturbance; TC, tropical cyclone; and -• -• -, convergence zone.

northeast current from the tropical disturbance of the NH near Africa and lasted 6 h. Other four MTCVs were formed in the flow of the southern hemisphere without the TC influence.

3.3.3 200/300 hPa

The principal influence of the NH flow was observed at the high levels (**Figure 5**, **Table 3**). Divergent flow was observed at the high levels in the TC. The height of this divergent flow depended on the TC intensity and was registered at 300 and 200 hPa. This divergent flow at the high levels passed through the equator to the BNE region. This flow from the TC was connected with anticyclonic circulation over the equatorial and south Atlantic. Thus, a single current was formed in the Northern Hemisphere, from the north or northeast.

Also, the anticyclonic circulation was predominant over the central and south region of the BNE. The location of this high was dependent upon the subtropical jet stream of the Southern Hemisphere (SJSHS). The northern boundary of this jet stream reached 15 or 30°W in the study events and was not observed over the BNE region. Curvature of the SJSHS was cyclonic (*trough*, *upper tropospheric trough*) over the continental region and anticyclonic (*ridge*) over the ocean region.

These two currents from the Northern Hemisphere and anticyclonic circulation of the Southern Hemisphere created the convergence of air current which, in turn, produced air sinking.

Two examples of this flow at the high levels can be seen in **Figures 5** and **6**. The difference between these examples is in the location of the Highs over the Atlantic Ocean. Meanwhile, the location of the convergence zone over the BNE was very similar.

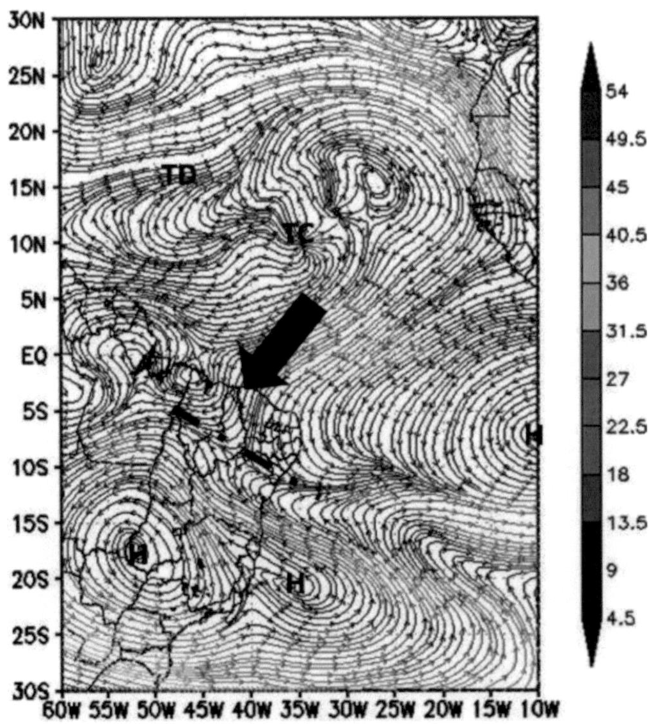

Figure 5.
Streamlines at 300 hPa 18 September 2015, 12UTC. H, anticyclonic circulation; TD, tropical disturbance; and TC, tropical cyclone. -•-•-, convergence zone.

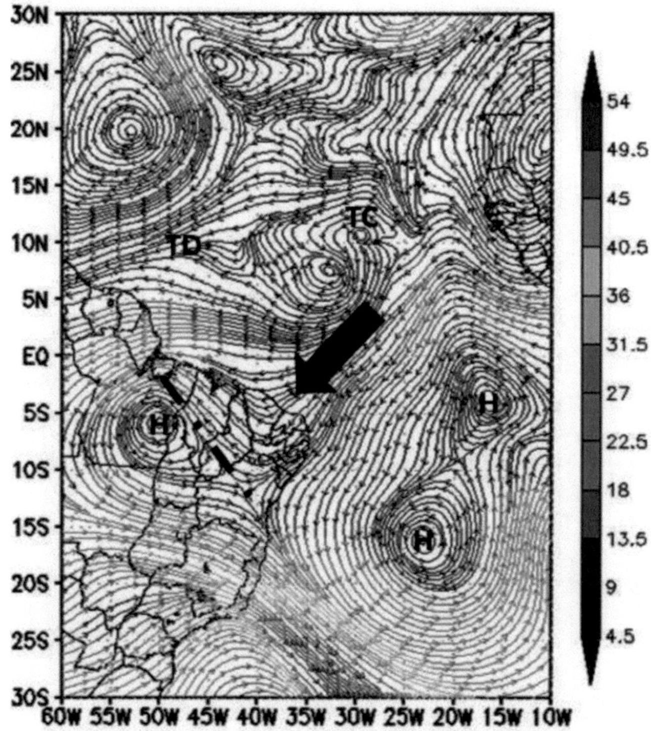

Figure 6.
Streamlines at 300 hPa 16 August 2013, 00UTC. H, *anticyclonic circulation;* TD, *tropical disturbance;* TC, *tropical cyclone; and -•-•-, convergence zone.*

A slightly different situation was observed during the TC Dorian on 25 July 2013 (**Figure 7**). This difference was associated with jet stream formation from the Northern to Southern Hemisphere, denominated as the Brazilian Northeast Jet Stream—BNEJS [20]. The data from this paper shows that, generally, BNEJS from the north was observed very rarely. On this day, the BNEJS was formed between two highs: in the north (near Maranhao State, high 1) and northeast of Brazil (high 2). The high 1 location was more to the north than the typical high position described before. At the same time, the high 2 location was close to the continental region, with its center on the northeastern cape of the South American continent. These highs squeezed the flow from the TC, and so the BNEJS was created. Also, the northern boundary of the SJSHS reached 15°W on this day. Therefore, the convergence zone between the currents from the south and north was detected in the south BNE region.

A scheme of the synoptic systems, observed more frequently at all levels simultaneously during the TC and fog events, is presented in **Figure 8**. This figure shows the circulation pattern, which contributes to fog, mist, and weak rain formation in the BNE. As a result of this circulation, a convergence zone at the middle and high levels and anticyclonic circulation at the middle levels were formed.

3.4 Mechanism of fog, mist, and weak rain formation

The circulation described in Section 3.3 creates sinking over the BNE. Vertical movements were analyzed in all events, and an example of this vertical motion is presented in **Figure 9**. Lifting in TC Bertha can be seen in **Figure 9a** and **d**. The vertical movement reaches 0.3 m/s. Very intense diffluent current was observed at

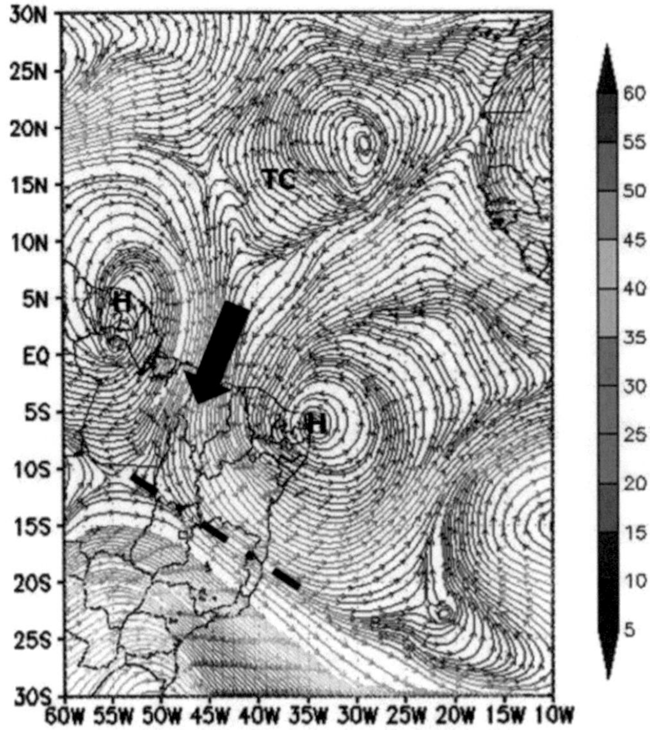

Figure 7.
Streamlines at 200 hPa, 25 July 2013, 06UTC. H, anticyclonic circulation; TC, tropical cyclone; and -•-•-, convergence zone. The black arrows show the BNEJS.

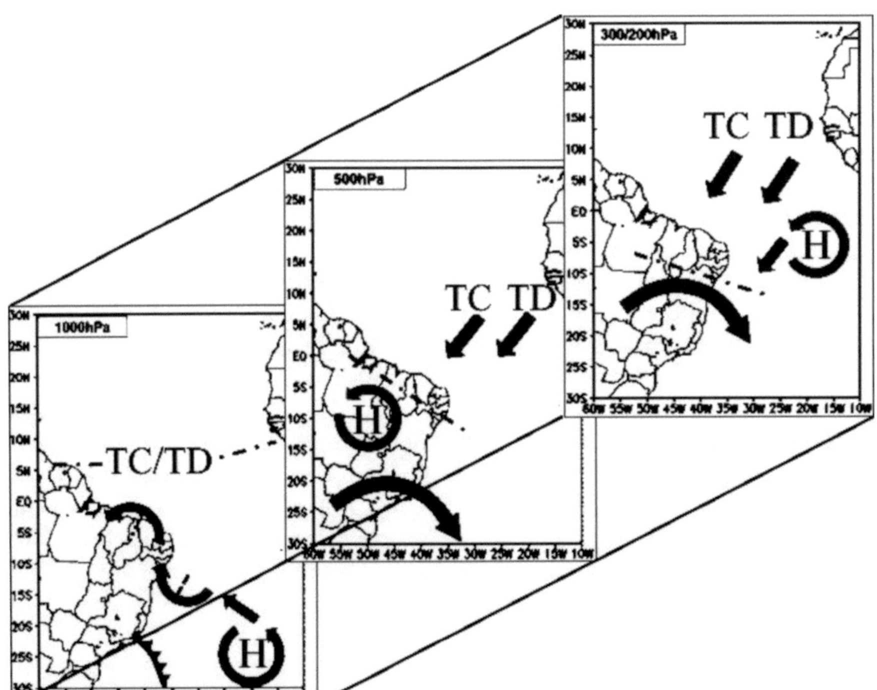

Figure 8.
Synoptic systems, observed more frequently at all levels simultaneously. TD, tropical disturbance; and TC, tropical cyclone.

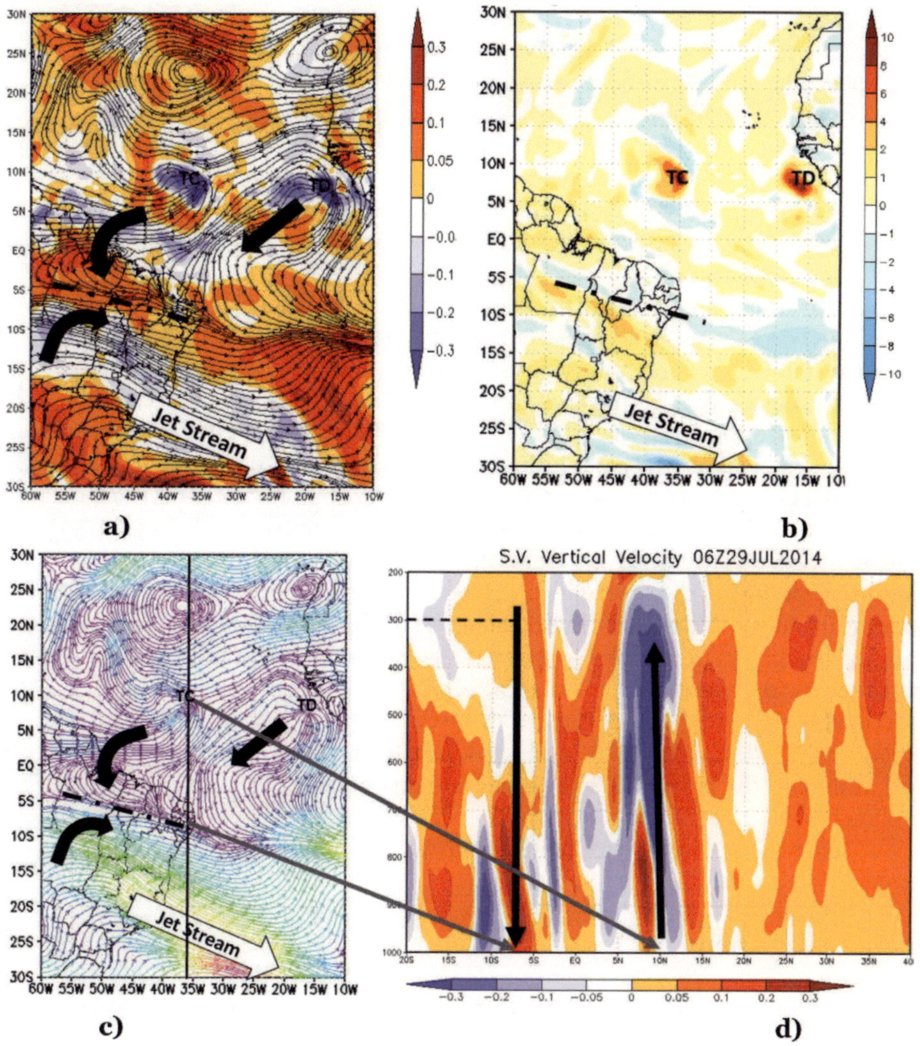

Figure 9.
Circulation between TC, TD is, and fog formation region on 29 June 2014, 06 UTC, in different maps: Vertical velocity (a, Pa s^{-1}) and divergence (b, s^{-1}), streamlines and wind velocity (c, m s^{-1}) all at 300 hPa and (d) vertical section of vertical velocity (Pa s^{-1}) along 37.1°W (longitude of TC Bertha).

300 hPa in TC Bertha (**Figure 9b**). Airflow at the high levels from TC Bertha to the BNE is presented in **Figure 9a** and **c**.

These circulations create sinking over the BNE and, as a result, humidity accumulation at the low levels and fog, mist, and weak rain formation. These phenomena were detected as the principal adverse phenomena in the BNE during all TC days.

4. Conclusion

The impact of all studied (10) TCs on the weather conditions in the BNE was analyzed during July to September from 2013 to 2015. These TCs had a *tropical storm* stage, when passing between 35 and 50°W and up to 20°N in the Northern Hemisphere. Only one, TC Bertha, had the stage of *tropical disturbance (TDis)* during its passage in this area.

The fog, mist, and weak rain in the BNE were the adverse phenomena associated more frequently with these TCs. Days with mist were also accompanied by weak rain. These phenomena were observed in all TC events and predominantly near the coastal region of the BNE.

The synoptic situation at 1000 hPa shows a wave disturbance near the BNE coastal area with a weak trough over the ocean close to the coastal area and a weak ridge over the northern region of the BNE. This situation was detected in all study events. Frontal extremity over the south of the Bahia State had an influence on trough intensification.

The convergence zone at the middle levels in the northern or middle region of the BNE was formed in all fog events between the currents from a TC in the Northern Hemisphere and anticyclonic circulation in the Southern Hemisphere. This anticyclonic circulation was predominant over the greater part of the BNE. Middle tropospheric cyclonic vortices were observed during half of the events but did not change the principal influence of the airflow from the TCs. One MTCV was formed between west flow of the Southern Hemisphere and northeast current from the tropical disturbance of the Northern Hemisphere.

The airflow at the high levels from the Northern Hemisphere is formed by the coupling (joining) of two currents: (1) from a TC and (2) from the high over the equatorial and south Atlantic. The confluence of these airflows with the current from the trough created by the subtropical jet stream of the Southern Hemisphere was the principal mechanism of the sinking in the BNE.

This sinking over the BNE formed the accumulation of humidity at the low levels and, as a result, fog, mist, and weak rain formation. Therefore, fog, mist, and weak rain were the principal adverse phenomena in the BNE during TC days.

This information can be used for short-term forecasting of adverse phenomena, such as fog and mist with weak rain in the BNE.

Acknowledgements

The authors thank English editor Robert Dower for his careful revision of this manuscript.

Author details

Natalia Fedorova*, Vladimir Levit and Lucas Carvalho Vieira Cavalcante
Institute of Atmospheric Science, Federal University of Alagoas, Brazil

*Address all correspondence to: nataliabras@gmail.com

References

[1] Fedorova N, Levit V. Fog in the Tropical Region. Fog Formation in the Tropical Region of the Northeast of Brazil. Saarbrücken, Germany: LAP LAMBERT Academic Publishing; 2016, 82 p. ISBN: 978-3-659-87098-9

[2] Schwerdtfeger W. Climates of Central and South America. New York: Elsevier Scientific Publishing Company; 1976

[3] Ratisbona LR. The climate of Brazil. Chapter 5. In: Schwerdtfeger W, editor. Climates of Central and South America. Oxford: Elsevier Scientific Publishing Company; 1976. pp. 219-269

[4] Silveira PS. Analysis of the fog cases and stratus clouds in Airport of Maceio [MSc thesis]. Maceio, Brazil: Federal University of Alagoas; 2003

[5] Fedorova N, Levit V, Fedorov D. Fog and stratus formation on the coast of Brazil. Atmospheric Research. 2008;**87**:268-278

[6] Fedorova N, Levit V, Souza JL, Silva AO, Afonso JMS, Teodoro I. Fog events at Maceio airport on the northern coast of Brazil during 2002-2005 and 2007. Pure and Applied Geophysics. 2015;**172**:2727-2749. DOI: 10.1007/s00024-014-1027-0

[7] Satyamurty P, Nobre CA, Silva Dias PL. South America. In: Karoly DJ, Vincent DG, editors. Meteorology of Southern Hemisphere. Boston: American Meteorological Society; 1998. pp. 119-139

[8] Molion LCB, Bernardo SO. Uma revisão da dinâmica das chuvas no Nordeste Brasileiro. Revista Brasileira de Meteorologia. 2002;**17**(1):1-10

[9] Rodrigues LRL, Fedorova N, Levit V. Adverse meteorological phenomena associated with low level baric troughs in the Alagoas State in 2003. Atmospheric Science Letters. 2010. DOI: 10.1002/asl.273

[10] Gomes HB, Fedorova N, Levit V. Rare events of stratus clouds on the northeast coast of Brazil. Revista Brasileira de Meteorologia. 2011;**26**(1):9-18

[11] Fedorova N, Levit V, Silva AO, Santos DMB. Low visibility formation and forecasting on the northern coast of Brazil. Pure and Applied Geophysics. 2013;**170**(4):689-709. DOI: 10.1007/s00024-012-0565-6

[12] Fedorova N, Levit V, Campos AMV. Brazilian northeast jet stream: Frequency, wind speed and direction. Meteorological Applications. 2018;**25**:254-260. DOI: 10.1002/met.1688

[13] Fedorova N, Levit V, Campos AMV. Brazilian northeast jet stream: Association with synoptic scale systems. Meteorological Applications. 2018;**25**:261-268. DOI: 10.1002/met.1693

[14] Fedorova N, Lyra MJA. Capítulo 1. Corrente de jato e fenômenos associados. In: Meteorologia em tópicos. Vol. 5. Pelotas: UFPel; 2017. pp. 11-68. ISBN 978-85-68891-04-9

[15] Gan MA, Kousky VE. Vórtices ciclônicos da Alta troposfera no Oceano Atlântico Sul (Cyclonic vortices in the upper troposphere the South Atlantic Ocean). Revista Brasileira de Meteorologia. 1986;**1**:19-28

[16] Kousky VE, Gan MA. Upper tropospheric cyclonic vortices in the tropical South Atlantic. Tellus. 1981;**33**(6):538-551

[17] Mishra SK, Rao VB, Franchito SH. Genesis of Northeast Brazil upper tropospheric cyclonic vortex: A primitive equation barotropic instability study. Journal of the Atmospheric Sciences. 2007;**64**:1379-1392

[18] Rao VB, Bonatti JP. On the origin of upper tropospheric cyclonic vortex in the South Atlantic Ocean and adjoin Brazil during summer. Meteorology and Atmospheric Physics. 1987;**37**:11-16

[19] Mishra SK, Rao VB, Gan MA. Structure and evolution of the large-scale flow of an embedded upper tropospheric cyclonic vortex over Northeast Brazil. Monthly Weather Review. 2001;**129**:1673-1688

[20] Fedorova N, Santos DMB, Lopes Segundo MM, Levit V. Middle tropospheric cyclonic vortex in northeastern Brazil and the tropical Atlantic. Pure and Applied Geophysics. 2017;**174**(1):397-411. DOI: 10.1007/s00024-016-1381-1

[21] Pontes da Silva BF, Fedorova N, Levit V, Peresetsky A. Sistemas sinóticos associados às precipitações intensas no Estado de Alagoas (Synoptic systems associated with the precipitations in the Alagoas State). Revista Brasileira de Meteorologia. 2011;**26**(3):295-310

[22] Fedorova N, Carvalho MH. Processos sinóticos em anos de La Niña e de El Niño. Parte II: Zonas Frontais. Revista Brasileira de Meteorologia. 2000;**15**(2):57-72

[23] Kousky VE. Frontal influences on Northeast Brazil. Monthly Weather Review. 1979;**107**(9):1140-1153

[24] Fedorova N, Levit V, Cruz CD. On frontal zone analysis in the tropical region of the Northeast Brazil. Pure and Applied Geophysics. 2016;**173**:1403-1421. DOI: 10.1007/s00024-015-1166-y

[25] Reeder MJ, Smith RK. Mesoscale meteorology. In: Meteorology of the Southern Hemisphere. Vol. 27(49). Boston: American Meteorological Society; 1998. pp. 201-241

[26] Fortune MA, Kousky VE. Two severe freezes in Brazil: Precursors and synoptic evolution. Monthly Weather Review. 1983;**111**:181-196

[27] Parmenter FC. A southern hemisphere cold-front passage near the equator. Bulletin of the American Meteorology Society. 1976;**57**:1435-1440

[28] Taljaard JJ. Synoptic meteorology of the Southern hemisphere. In: Meteorology of the Southern Hemisphere. Vol. 13(35). Boston: Press AMS; 1972. pp. 129-213

[29] Hastenrath S. Climate Dynamics of the Tropics. Dordrecht: Kluwer Academic Publishers; 1991. p. 488

[30] Vasquez T. Weather Forecasting Handbook. Garland, Texas: Weather Graphics Technologies; 2000. 98 p

Chapter 3

A Recurrence Analysis of Multiple African Easterly Waves during Summer 2006

Tiffany Reyes and Bo-Wen Shen

Abstract

Accurate detection of large-scale atmospheric tropical waves, such as African easterly waves (AEWs), may help extend lead times for predicting tropical cyclone (TC) genesis. Since observed AEWs have comparable but slightly different periods showing spatial and temporal variations, local analysis of frequencies and amplitudes of AEWs is crucial for revealing the role of AEWs in the modulation of TC genesis. To achieve this goal, we investigate the recurrence plot (RP) method. A recurrence is defined when the trajectory of a state returns to the neighborhood of a previously visited state. To verify implementation of the RP method in Python and its capability for revealing a transition between different types of solutions, we apply the RP to analyze several idealized solutions, including periodic, quasiperiodic, chaotic and limit cycle solutions, and various types of solutions within the three- and five-dimensional Lorenz models. We then extend the RP analysis to two datasets from the European Centre for Medium-Range Weather Forecasts global reanalysis and global mesoscale model data in order to reveal the recurrence of multiple AEWs during summer 2006. Our results indicate that the RP analysis effectively displays the major features of time-varying oscillations and the growing or decaying amplitudes of multiple AEWs.

Keywords: recurrence plot, chaos, limit cycle, AEWs, Lorenz model

1. Introduction

Recent studies suggest that accurate detection of recurrent, multiple, large-scale tropical waves, such as African easterly waves (AEWs), has the potential to help extend prediction lead times for tropical cyclone (TC) genesis. For example, using the NASA global mesoscale model (GMM, e.g., [1]), the formation, subsequent intensification, and movement of hurricane Helene (2006) were simulated to a degree of satisfaction from Day 22 to 30 within a 30-day model integration when multiple AEWs were realistically simulated (e.g., [2]). To date, the scalable parallel ensemble empirical mode decomposition method (PEEMD; e.g., [3, 4]) has revealed the role of downscaling processes associated with environmental flows in determining the timing and location of hurricane Helene's formation (e.g., [5]), supporting the model's practical predictability over extended-range time scales. Prior to Helene's formation, observed AEWs had comparable but slightly different periods with both spatial and temporal variations. Thus, a local analysis of the

IntechOpen

frequencies and amplitudes for multiple AEWs is desired in order to monitor the evolution of AEWs that may influence the timing and location of TC genesis.

To effectively reveal space-varying features and/or temporal oscillations in simulations and reanalysis data, recurrence plots (RPs) are employed (e.g., [6]). Recurrence is defined when the trajectory of a state returns back to the neighborhood of a previously visited state. Thus, recurrence may be viewed as a generalization of periodicity to brace quasiperiodicity [7] and chaos [8]. Measuring the patterns associated with recurrence provides valuable information regarding the oscillatory nature of a system. To verify implementation in Python, we test the RP method using several types of idealized solutions and three types of solutions within the three- and five-dimensional Lorenz model (3DLM and 5DLM). We then apply the method in order to analyze two global datasets.

The paper is organized as follows. Section 2.1 introduces the European Centre for Medium-Range Weather Forecasts (ECMWF) global reanalysis data and NASA GMM data (e.g., [1, 2, 4, 9]). We focus on a 30-day period in 2006 over which multiple AEWs were observed during the NASA African Monsoon Multidisciplinary Analyses (NAMMA) field campaign (e.g., [2, 10]). Section 2.2 introduces the 3DLM and 5DLM (e.g., [11–13]). Section 2.3 introduces RP methods for performing the local analysis. In Section 3.1, the method is applied in order to analyze four basic types of solutions, including (1) periodic, (2) quasiperiodic, (3) chaotic solutions, and (4) Gaussian white noise. While Section 3.2 presents an analysis of the idealized spiral sink, Section 3.3 extends the RP analysis for steady-state solutions of the 3DLM. Section 3.4 discusses limit cycle solutions of the 3DLM and 5DLM and their RP analysis. Section 3.5 presents an analysis of global reanalysis and GMM data in order to reveal the time-varying recurrent behavior of AEWs. Appendix A discusses the mathematical approach for the unique version of the 3DLM and 5DLM, referred to as the Version 2 (V2) system. V2 systems are used for either linear or nonlinear simulations. Additionally, the linearized V2 system is applied to perform an eigenvalue analysis to obtain the growth rates and oscillatory frequencies of solutions.

2. Global data, the Lorenz model, and the recurrence analysis method

2.1 Global reanalysis and model data

For this study, we select the latest European Centre for Medium-Range Weather Forecasts (ECMWF) global reanalysis (ERA-interim) dataset. The data spans the time period from January 1979 to the present day. The dataset has a sampling rate of 6 h and a horizontal grid spacing of 0.75°, yielding a spatial resolution of approximately 78 km. The following gridded products are available: 3-h surface fields, daily vertical integrations, 6-h upper air atmospheric fields for pressure, potential temperature, potential vorticity levels, and vertical coverage from the lower troposphere to the stratosphere on 60 model layers. Based on previous studies (e.g., [2, 5]), here, we focus on the 30-day period from August 22 to September 20, 2006. During this period, the NASA African Monsoon Multidisciplinary Analyses (NAMMA) field campaign documented multiple AEWs and provided a great opportunity to characterize the frequency of AEWs as well as their evolution and impact over continental western Africa. **Figure 1a** and **b** display multiple AEWs that were detected by a change in sign of meridional winds from the ERA-interim dataset and the NAMMA observations, respectively.

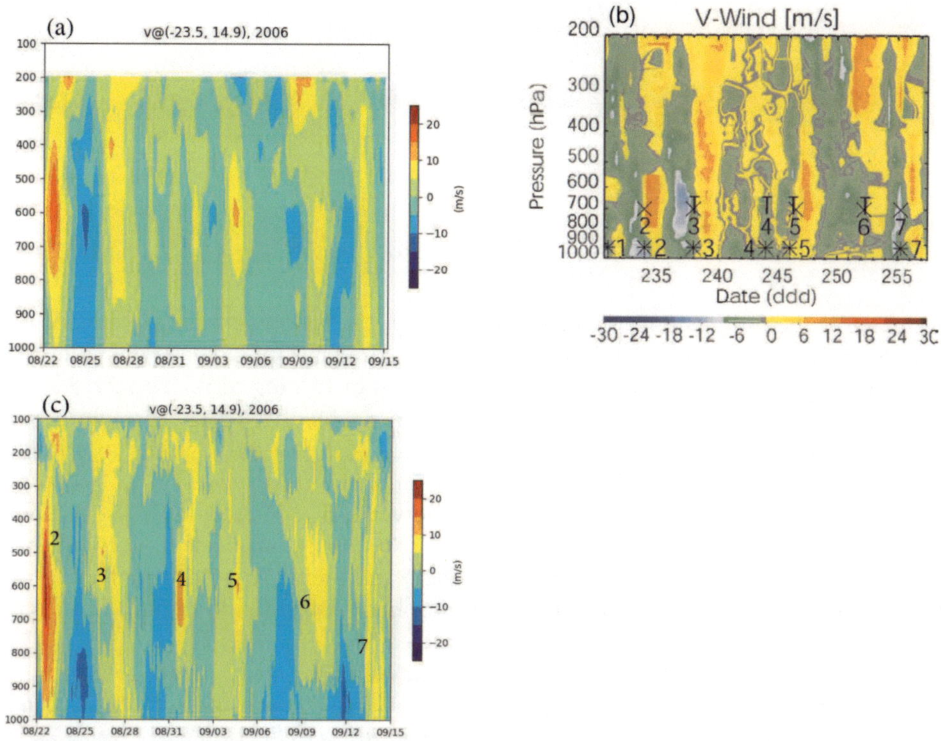

Figure 1.
A time-altitude cross section of meridional winds from the ERA-interim dataset (a) and global mesoscale model simulations (c). The color bar to the right of each figure represents the wind velocity in meters per second. Panel (b) displays NAMMA observations for a comparison (e.g., [10]).

As a result of higher spatial and temporal resolutions, GMM simulation data [1, 9] are also selected. The GMM that produced the dataset is composed of three major components: finite volume dynamics [14], National Center for Atmospheric Research (NCAR) Community Climate Model physics, and the NCAR Community Land Model. Simulations spanning the 35 day period began on August 22, 2006. However, our analysis only focuses on the first 30 days. The data set has a 15-min integration time step, 17 vertical levels, and a horizontal resolution of approximately 28 km. **Figure 1c** provides the time-altitude cross section of meridional winds from the GMM dataset.

2.2 The five-dimensional Lorenz model (5DLM)

In the real world, as a result of nonlinearity, system solutions may possess time-varying frequencies and amplitudes. To understand the performance of the selected method in analyzing nonlinear time series data, the 3DLM and 5DLM are used to provide three types of solutions at different ranges of parameters for verifications. The 5DLM, which extends the 3DLM and includes the 3DLM as a subset (e.g., [13]), is defined by Eqs. (1)–(5), as follows:

$$\frac{dX}{d\tau} = -\sigma X + \sigma Y, \tag{1}$$

$$\frac{dY}{d\tau} = -XZ + rX - Y, \tag{2}$$

$$\frac{dZ}{d\tau} = XY - XY_1 - bZ, \tag{3}$$

$$\frac{dY_1}{d\tau} = XZ - 2XZ_1 - d_o Y_1, \tag{4}$$

$$\frac{dZ_1}{d\tau} = 2XY_1 - 4bZ_1. \tag{5}$$

Here, τ is the dimensionless time. In this study, the following parameters are kept constant: $b = 8/3$, $d = 19/3$, and $\sigma = 10$. By comparison, we vary the value of the Rayleigh parameter, r, that represents a heating or forcing term. The term XY_1 in Eq. (3) plays a role in providing negative nonlinear feedback associated with small-scale dissipative processes (e.g., $-d_o Y_1$ and $-4bZ_1$) to stabilize the solutions (e.g., [13]). Equations. (1)–(3), without inclusion of the feedback term (XY_1), are reduced to become the 3DLM. As discussed below, the 5DLM requires higher values of r for the onset of chaos (e.g., [13]). Recently, the 5DLM has been re-derived and analyzed in detail by Moon et al. [15] and Felicio and Rech [16], showing the model's robustness. Furthermore, the 5DLM has been extended to higher-dimensional LMs (e.g., [17–20]) and a generalized LM (e.g., [21, 22]).

The 3DLM produces three types of solutions, including steady-state solutions, chaotic solutions, and nonlinear periodic solutions. We present the first two types of solutions below and discuss the third type of solution in Section 3.4. **Figure 2a** provides Y and Z modes of the solution for the strange attractor using the 3DLM with $r = 25$. Irregular oscillations surrounding nontrivial critical points that are defined in Appendix A are indicative of chaotic solutions. In **Figure 2b**, the 5DLM with $r = 25$ clearly produces a steady-state solution that spirals into the non-trivial critical point. **Figure 2c** from the 5DLM with $r = 43.5$ appears more like **Figure 2a** and depicts an irregular oscillatory motion. Such behavior within the 5DLM only occurs when $r \geq r_c = 42.9$ and when the other parameters are held at the aforementioned values. The corresponding solutions for Y1 and Z1 are shown in **Figure 2d**.

2.3 Recurrence plot analysis

The recurrent nature of a system can offer important insight into its dynamics, such as periodicities or other oscillatory behavior (e.g., growing or decaying oscillations). In the case of deterministic systems, the recurrence of states (i.e., a state returning to an arbitrarily close area at a point from a previous time within the phase space) is an important feature (e.g., [6]). Such recurrences can be visualized using a tool known as recurrence plots (RPs). RPs allow for the representation of multidimensional systems in a two-dimensional visualization; thus, they easily provide insight into high-dimensional systems. The RP plots a black point for each time (i,j) in which there is a recurrence, more precisely given by the following equation:

$$R_{i,j} = \Theta\left(\varepsilon - \|\vec{x_i} - \vec{x_j}\|\right) \quad i,j = 1, \ldots, N, \tag{6}$$

where N is the number of states ($\vec{x_i}$) considered, ε is the threshold distance for recurrence, $\|\cdot\|$ is the Euclidean norm, and $\Theta(\cdot)$ is the Heaviside function (e.g., [23]). The individual point (i,j) does not offer any information regarding the states of the system at times i and j. However, the phase space trajectory can be reconstructed from the collection of all of the recurrence points present within

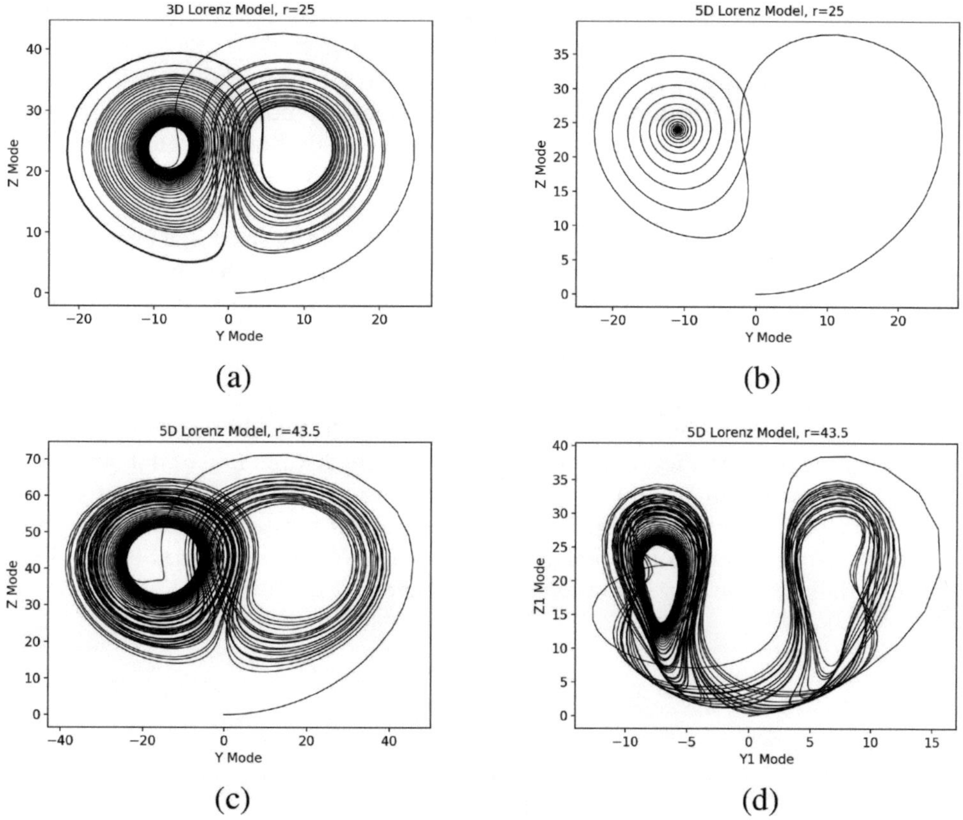

Figure 2.
Phase space plots for the 3DLM (a) and 5DLM (b–d). All of the plots show Lorenz strange attractors. (a) Y *vs.* Z *modes using the 3DLM with* $r = 25$*. (b and c)* Y *vs.* Z *modes using the 5DLM with* $r = 25$ *and* $r = 43.5$*, respectively. (d)* Y_1 *vs.* Z_1 *modes with* $r = 43.5$ *(see details in [13]).*

the RP (e.g., [6]). As shown in **Figure 3a**, a typical RP for a simple periodic solution displays uniform diagonal lines. Here, it should be noted that the RP has a main black diagonal line with an angle of $\pi/4$, referred to as the line of identity (LOI).The threshold distance for recurrence (ε) is an important parameter to consider. If the value is too small, there may not be enough recurrences plotted, resulting in insufficient information required for conducting an accurate analysis. For a value that is too large, the appearance of many recurrent points may mask important structures that would otherwise be present within the RP. While there is much discussion regarding the proper selection of ε, a general rule is to find the smallest possible value that is capable of providing a sufficient number of recurrences for identifying recurrent structures within a dataset (e.g., [6]). Figure 2.10 of Reyes [24] demonstrates the impact with different values of ε on the RP for the 3DLM.

3. Results

In this section, we first discuss the RP analysis for idealized solutions, various types of solutions from the 3DLM and 5DLM. We then apply the RP to analyze the global reanalysis and the GMM 30-day simulation data in order to detect recurrent, multiple AEWs that play an important role in the modulation of TC genesis (e.g., Hurricane Helene (2006)).

3.1 Analysis of four basic types of solutions

Figure 3 provides RPs for four types of solutions, including (a) periodic solutions that are characterized by uniform diagonal lines extending from edge to edge of the plot and spaced with an equal distance from one another; (b) quasiperiodic solutions with a similar structure to the periodic RP, with the exception that the distance between the diagonal lines varies; (c) chaotic solutions from the 3DLM with a Rayleigh parameter of $r = 28$ whose plots have many short diagonal lines spaced various distances from one another; and (d) Gaussian white noise, characterized by a homogenous RP consisting of many individual points and little to no organized recurrent structures (e.g., [6]).

Other features of the RP analysis, as discussed below, include (1) vertical or horizontal lines or clusters, which indicate small or no change in solution amplitudes over a certain period of time, and (2) white bands, which suggest transitions from one type of dynamics to another. More than one distinct line or dot structure may also appear within a single RP. Line structures and the length and number of lines, as well as their positioning, describe the dynamics for each system (e.g., [25]). When applying this method to real data, the overall structures should still be visible, although they may or may not be as neatly depicted.

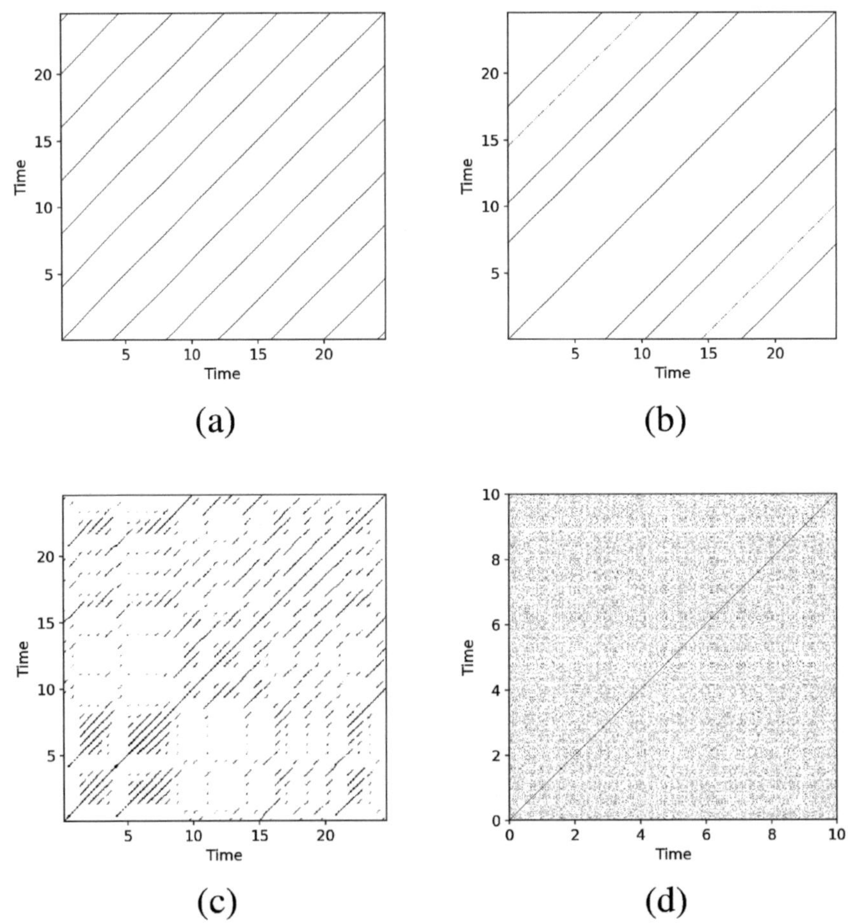

Figure 3.
Recurrence plots of (a) periodic solutions produced using $y = sin\,(2\pi t/4)$*, (b) quasiperiodic solutions produced using* $y = sin\,(16\pi t) + sin\,(8\sqrt{(2)}\pi t)$*, (c) chaotic solutions produced by the 3DLM, and (d) Gaussian white noise.*

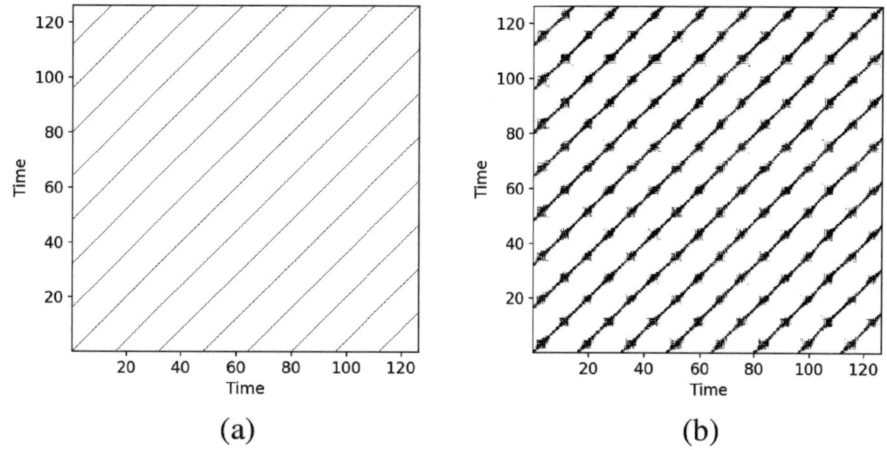

Figure 4.
Recurrence plots of raw data with a periodic solution given by $y = \sin(2\pi t/16)$ *(a) and the raw data superimposed with Gaussian white noise that has a finite amplitude of 0.1 (b).*

Previously, performance of the RP was demonstrated using idealized data. In practice, most real data has some amount of random noise. Therefore, understanding the capabilities of the method in analyzing noisy data, before applying them to real or non-idealized datasets, is important. Here, we present an analysis of composite data consisting of the raw data and additional noise. **Figure 4a** displays the RP for idealized periodic solutions generated from $y = \sin(2\pi t/16)$, $0 \leq t \leq 128$, and $\Delta t = 0.2$. **Figure 4b** provides the RP for the same dataset superimposed with Gaussian white noise with an amplitude of 0.1. While the overall diagonal line structure is still present, some visual differences are apparent. The most prominent differences include thicker diagonal lines and a fluctuation in thickness. The differences are due to so-called false recurrence resulting from the noise itself and the slightly higher value of ε required in order to obtain a reasonable plot.

3.2 Analysis of a spiral sink and source

In addition to the four fundamental solutions in **Figure 3**, real-world model data may include growing or decaying oscillatory solutions. Mathematically, such a solution can be represented as $y = e^{\alpha t}\sin(\beta t + \theta)$, where θ is a phase constant. This type of solution can be obtained from a linearized system with a complex eigenvalue, $\lambda = \alpha + i\beta$, where α and β are real numbers. Next, we discuss the conditions under which a recurrence plot can be defined for the general oscillatory solution. To facilitate discussions, we further assume $\beta > 0$ and consider three cases with (i) $\alpha = 0$ for a simple oscillation, (ii) $\alpha > 0$ for a growing oscillation (i.e., a spiral source), and (iii) $\alpha < 0$ for a decaying oscillation (i.e., a spiral sink), respectively.

RPs for spiral sinks or sources display a different structure as compared to RPs for periodic and quasiperiodic solutions, as discussed below. **Figure 5a** displays an oscillatory solution, $y = e^{-0.1t}\sin(t)$ (i.e., $(\alpha, \beta, \theta) = (-0.1, 1, 0)$). The solution decays with time over the time interval from 0 to 128. When the solutions produce time-varying amplitudes, the RPs display vertical and horizontal lines. As the solution decays toward zero, the lines become denser (the upper right-hand corner of **Figure 5c**). Namely, the density of the lines in these RPs increases when the solution is closer to the mean value (i.e., zero for this idealized solution). The appearance of horizontal and vertical lines is explained below.

Here, we provide an additional analysis for the pattern of the RP in **Figure 5**. As shown in **Figure 6**, a distance (D) between two points $y(t_1)$ and $y(t_2)$ with a time lag of $T = \frac{2\pi}{\beta}$ is given by $D = |y(t_2) - y(t_1)|$, where $t_2 = t_1 + T$. We can obtain the following equation:

$$D = |e^{\alpha t_2} \sin(\beta t_2 + \theta) - e^{\alpha t_1} \sin(\beta t_1 + \theta)| \tag{7}$$

$$\begin{aligned} &= \left| e^{\alpha\left(t_1 + \frac{2\pi}{\beta}\right)} \sin\left(\beta\left(t_1 + \frac{2\pi}{\beta}\right) + \theta\right) - e^{\alpha t_1} \sin(\beta t_1 + \theta) \right| \\ &= \left| e^{\frac{2\alpha\pi}{\beta}} e^{\alpha t_1} \sin(\beta t_1 + \theta) - e^{\alpha t_1} \sin(\beta t_1 + \theta) \right| \\ &= \left| e^{\frac{2\alpha\pi}{\beta}} e^{\alpha t_1} \sin(\beta t_1 + \theta) - e^{\alpha t_1} \sin(\beta t_1 + \theta) \right| \\ &= \left| e^{\alpha t_1} \sin(\beta t_1 + \theta) \left(e^{\frac{2\alpha\pi}{\beta}} - 1 \right) \right|. \end{aligned} \tag{8}$$

Recurrence appears when the distance, D, is small (i.e., $D < \varepsilon$, where ε represents a threshold), requiring

$$\left| e^{\alpha t_1} \sin(\beta t_1 + \theta) \left(e^{\frac{2\alpha\pi}{\beta}} - 1 \right) \right| < \varepsilon. \tag{9}$$

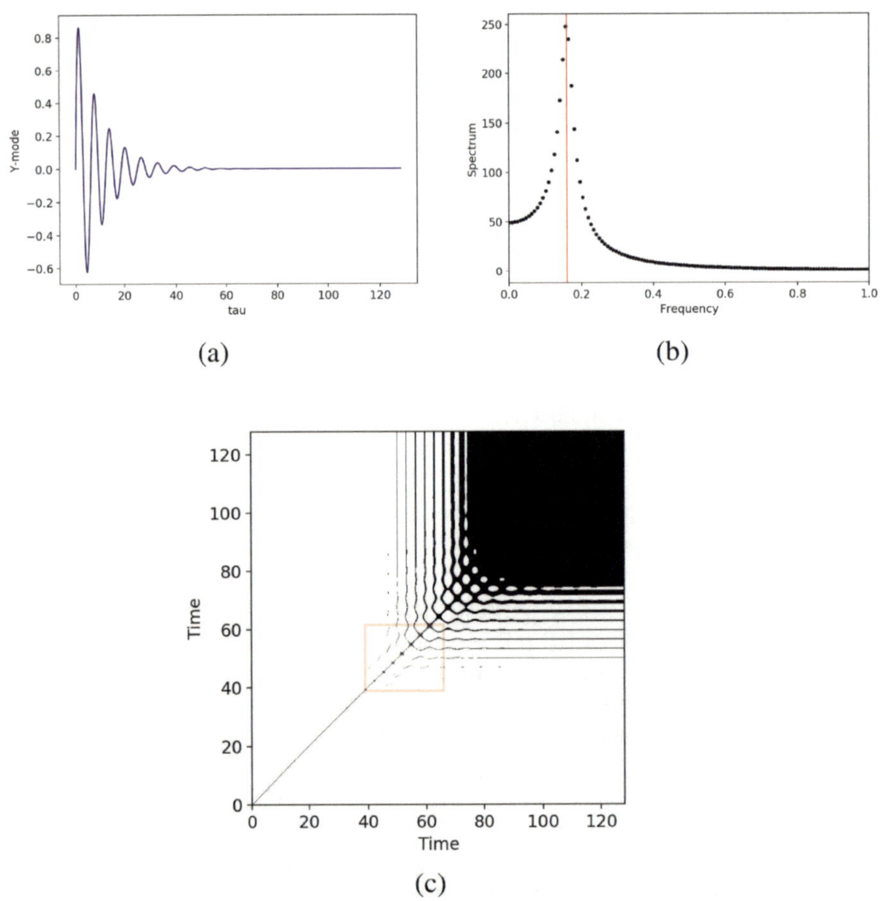

Figure 5.
A time evolution of the solution $y = e^{-01.t} \sin(t)$ (a), its power spectrum (b), and RP (c) for $\tau \in [0, 128]$.

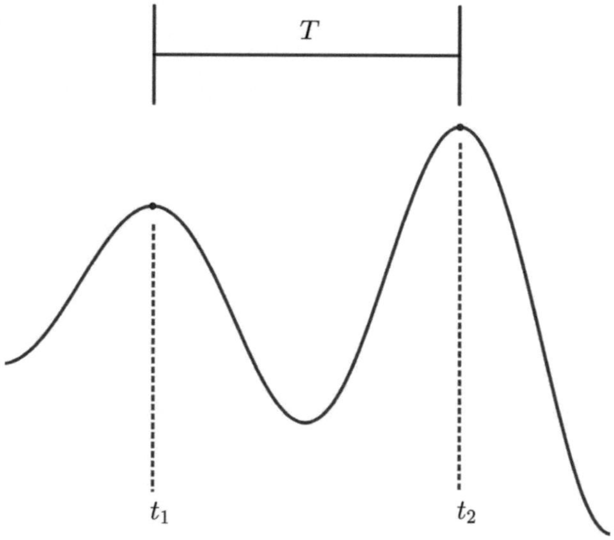

Figure 6.
An oscillatory mode and two points at t_1 and t_2, where $t_2 = t_1 + T$.

Therefore, (i) for a simple oscillation, $\alpha = 0$, Eq. (8) is always valid, suggesting that two points with a time lag of T define a recurrent point within the recurrence plot. For a specific t_1 where $\sin(\beta t_1 + \theta) = 1/M \neq 0$, Eq. (8) leads to:

$$|e^{\alpha t_1}| < \left| \frac{M}{\left(e^{\frac{2\alpha\pi}{\beta}} - 1\right)} \right| \varepsilon. \tag{10}$$

The value of αt_1 roughly determines if the above inequality is valid. As a result, Eq. (10) suggests that (ii) for a spiral source, $\alpha > 0$, a recurrence point appears when t_1 is small and that (iii) for a spiral sink, $\alpha < 0$, a recurrence point appears when t_1 is large. This type of RP appears in the red box in **Figure 5c**. Additionally, (iv) when $\sin(\beta t_1 + \theta) = 0$ in Eq. (7a), the point at $t = t_1$ can produce recurrence with all of the points at t_2 that have very small amplitudes (i.e., the amplitudes are close to zero). Thus, as shown in **Figure 5c**, the appearance of continuous horizontal lines indicates a transition from the third type of solution (i.e., a spiral sink) into the 4th type of solution with a small or zero amplitude. Note that the distance of two horizontal lines determined in (iv) yields an estimate of a half of the period (i.e., $T/2$.) In general, Eq. (9) suggests that the aforementioned recurrence threshold should be selected by taking the rate of growth or decay of the oscillatory solutions into account, which will be the subject of a future study.

3.3 Analysis of a steady-state solution within the 3DLM

While the 3DLM with $r > 24.74$ produces chaotic solutions, it simulates steady-state solutions when $r < 24.74$. Since the steady-state solution displays a decaying oscillatory mode, which may resemble a weakening AEW, its RP is analyzed below. **Figure 7a** displays numerical solutions of the Y mode of the 3DLM with $r = 10$. At approximately $t = 7.0$, only small fluctuations in the solution remain (i.e., the solution has a small amplitude over this interval). **Figure 7b** displays the corresponding recurrence plot with $\varepsilon = 0.05$. Initially, horizontal and vertical lines appear, consistent with the previous analysis using **Figures 5** and **6**. As the solution begins to converge to a steady state, the plot rapidly becomes denser.

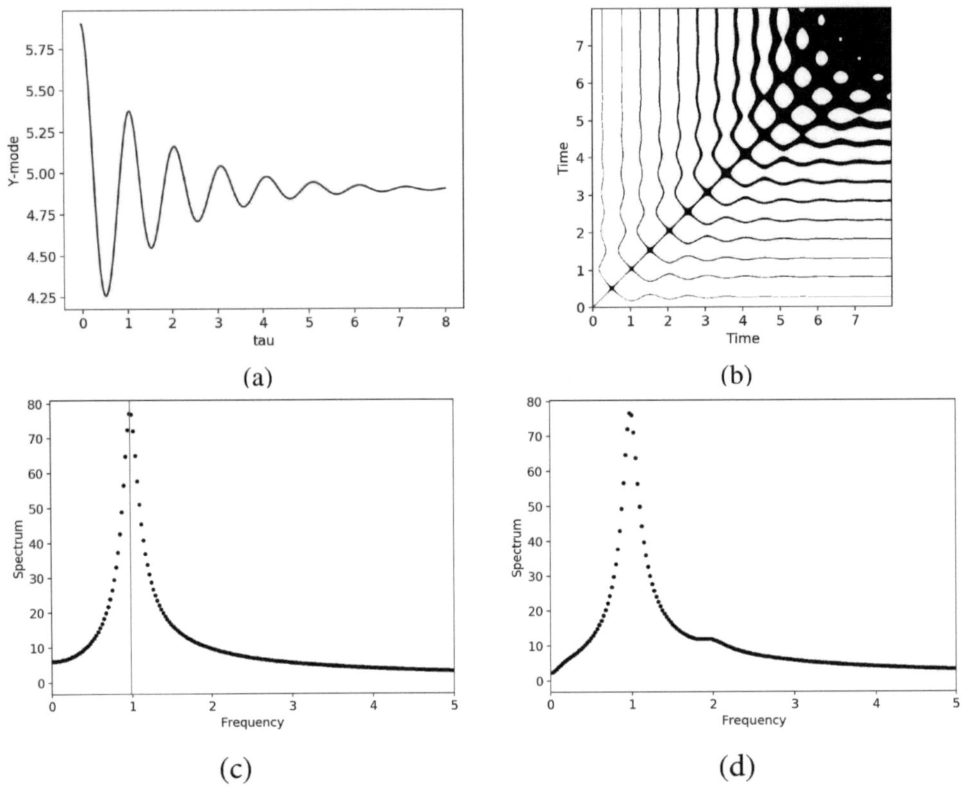

Figure 7.
Solutions for the Y mode of the 3DLM with $r = 10$ (a) and the corresponding recurrence plot (b). Power spectra for Y mode solutions of the 3DLM V2, $FN = 0$ (c) and $FN = 1$ (d).

By comparison, **Figure 7c** displays the power spectrum for the linear 3DLM V2 ($FN = 0$) with the same parameter values as the solution in **Figure 7a**. Eigenvalues for the 3DLM V2 $FN = 0$ are approximately $-0.59550 \pm 6.17416i,\ -12.47567$, indicating that the frequency of a linear solution near the nontrivial critical point is $6.174161/2\pi = 0.9826$, as confirmed by the power spectrum. However, such a spectral analysis does not reveal a local transition from a decaying oscillation to a constant solution. **Figure 7d** provides the power spectrum for the corresponding nonlinear simulations with $FN = 1$ within the 3DLM V2, which produces a result similar to panel (c). Comparable spectra for linear and nonlinear steady-state solutions are due to the fact that perturbations (which measure departures from the nontrivial critical point) become smaller as time proceeds when solutions move closer to the critical point.

3.4 Analysis of limit cycle solutions

At moderate heating parameters, both the 3DLM and 5DLM models produce chaotic solutions when other parameters are held constant. By comparison, for very large values of r, the 3DLM and 5DLM produce limit cycle solutions. A limit cycle is defined as a closed and isolated orbit to which nearby trajectories converge (e.g., [26, 27]). In **Figures 8** and **9**, limit cycle solutions are simulated in order to compare the 3DLM V2 and 5DLM V2 with the original 3DLM and 5DLM with $r = 800$. In all three panels for both plots, only the trajectory in red is seen, indicating that the V2 with $FN = 1$ and the original versions produce the same solutions over the given time intervals.

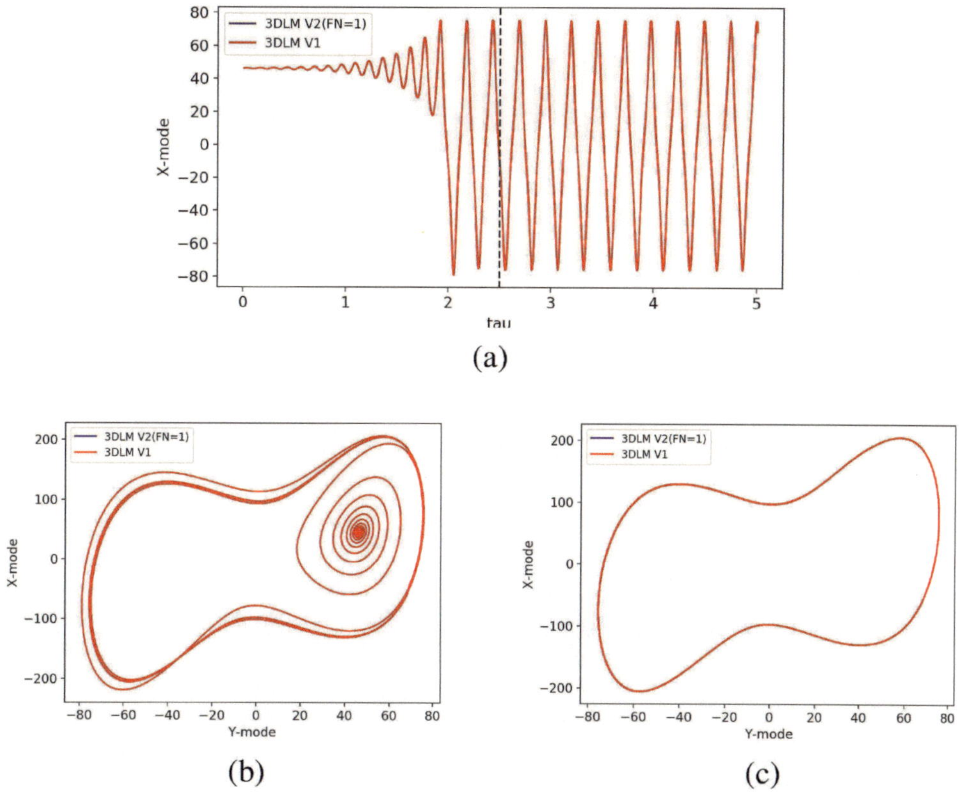

Figure 8.
Time evolution of X mode solutions (a) and X vs. Y mode solutions (b), (c) of the 3DLM V2 (FN = 1) and the original 3DLM with r = 800. Panels (a) and (b) display solutions over the time interval [0, 5], while panel (c) shows the solution over the interval [2.5, 5] during which the orbit becomes regularly oscillatory in (a).

In **Figure 8** for the 3DLM, we plot the time evolution of the X mode in panel (a). As shown, the solution begins by an oscillation with small time-varying amplitudes and gradually increases to become regularly oscillatory at approximately $\tau = 2.1$. In the $X - Y$ phase space in panel (b), the solution spirals away from the corresponding nontrivial critical point and then becomes a closed curve, leading to a limit cycle that encloses trivial and nontrivial critical points. The closed and isolated features of the limit cycle are clearly shown in panel (c) over the time interval $\tau \in [2.5, 5]$, displaying only one trajectory.

For comparison, a limit cycle solution of the 5DLM is plotted in **Figure 9**. Just as in plots for the 3DLM, the X mode of the 5DLM grows until it becomes oscillatory with a nearly constant amplitude after $\tau = 3.2$. **Figure 9b** plots solutions of the X and Y modes over $\tau \in [0, 8]$. The solution spirals outward from its initial position and gradually forms a limit cycle. Panel (c) displays the solution in the $X - Y$ space over $\tau \in [4, 8]$, showing a closed curve. The above spiral orbit and limit cycle and their transition are analyzed below using the RP.

Recurrence plots for limit cycle solutions exemplify the method's capacity for local analysis. **Figure 10a** and **b** provides RPs for the 3DLM and 5DLM with $r = 800$. In both of these RPs, four distinct line patterns can be sequentially identified: (1) a dark block or a cluster of points (bottom left), (2) vertical and horizontal lines, (3) a large white band or a lack of points (middle), and (4) uniformly spaced diagonal lines (upper right). As shown in **Figures 8** and **9**, these patterns correspond to a transition from a spiral solution near a nontrivial critical point to a (nonlinear) periodic solution that encloses trivial and nontrivial critical points.

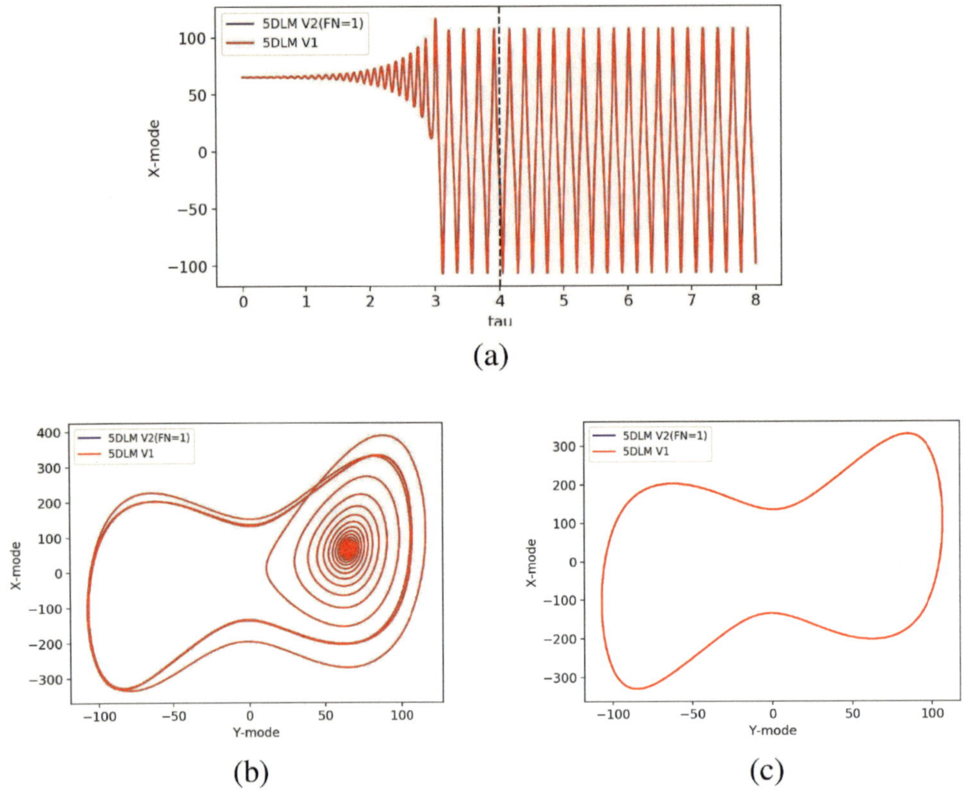

Figure 9.
Time evolution of X mode solutions (a) and X vs. Y mode solutions (b), (c) of the 5DLM V2 (FN = 1) and the original 5DLM with r = 800. Panels (a) and (b) show solutions over the time interval [0, 8], while panel (c) occurs over the interval [4, 8] during which the solution becomes a limit cycle.

Specifically, the solutions initially oscillate with very small amplitudes, progress into oscillations with larger amplitudes, and finally turn into regular oscillations. Since the patterns of the RPs directly correspond to a change in solution types, the RPs are capable of clearly identifying local dynamics and transition. By comparison, a spectral analysis cannot reveal such a local transition.

3.5 Analysis of global reanalysis and model data

Here, we apply the RP to analyze the case with multiple AEWs. **Figure 11** plots the time series for wind velocity at the 850 hPa level for the ERA-interim data (a) and GMM simulations (b). In panel (a), the amplitude of the signal is nearly constant over the entire duration. In panel (b), the signal's amplitude generally decays first and then grows with time. In comparison to the observations from the NAMMA field campaign (Figure 1b or Figure 4a of [2]), previous studies (e.g., [2, 28]) have indicated that the GMM simulations more accurately represent actual observations as compared to the reanalysis data.

Figure 11 displays the AEWs from the ERA-interim and GMM data, to be analyzed below. We first create two sets of idealized oscillatory solutions that mimic the "observed" and simulated AEWs. **Figure 12a** displays an idealized solution consisting of a periodic solution with a frequency of 1 and a periodic solution with a frequency of 2, separated by periodic solutions with negligible amplitude, similar to the ERA-interim data. In this simplified case, the oscillatory solutions both have the same amplitude. The corresponding power spectrum indicates peaks

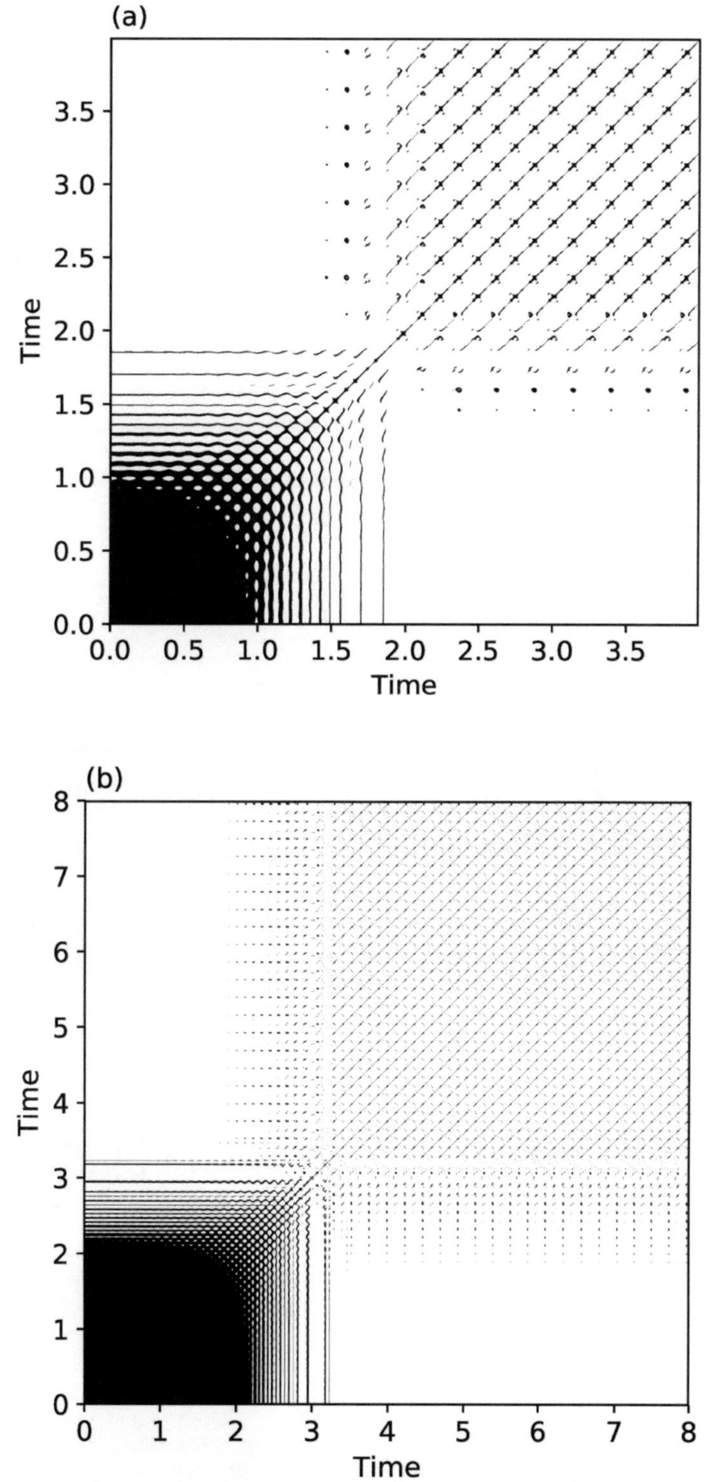

Figure 10.
A recurrence plot of a transition from a spiral orbit (the bottom left corner) to a limit cycle (the top right corner) within the 3DLM (top) and 5DLM (bottom) using $r = 800$.

at 1 and 2 but does not readily offer information regarding the wave transitions (not shown). The RP of the signal, as seen in **Figure 12b**, displays uniform diagonal lines in each corner and a solid block of recurrent points in the center. These features

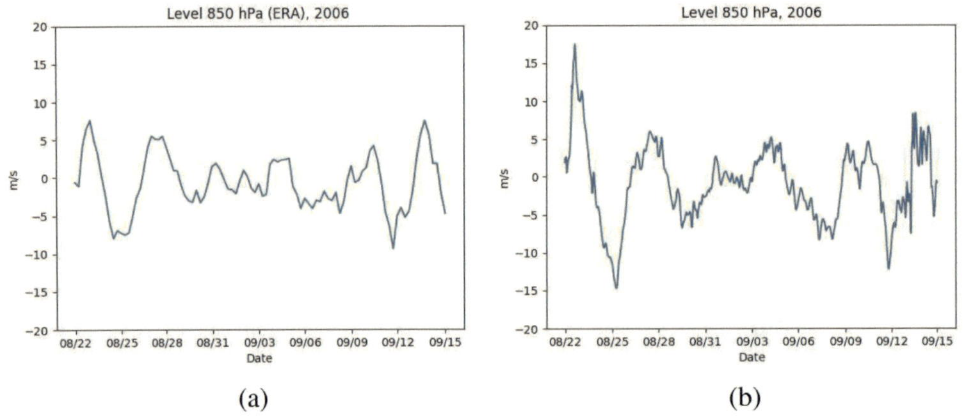

Figure 11.
Time series of the wind velocity at the 850 hPa level for the ERA-interim dataset (a) and the global mesoscale model (b).

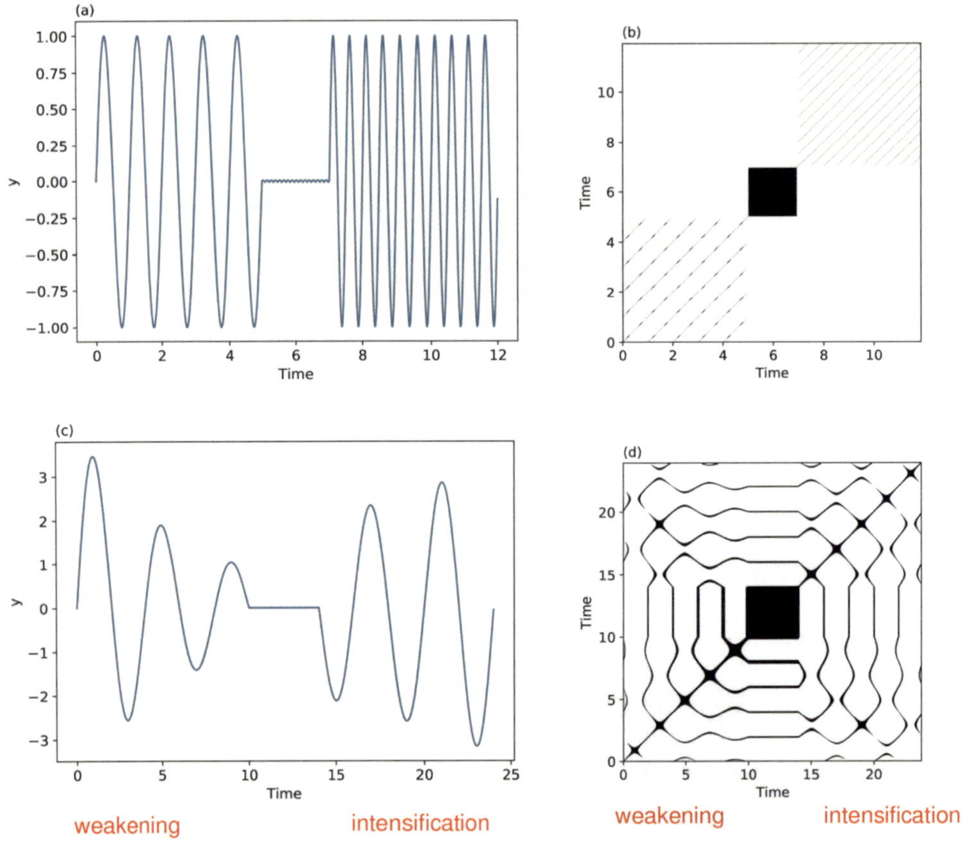

Figure 12.
Idealized oscillatory data used to mimic multiple AEWs with two different periods (a) and decaying and growing amplitudes (c). The corresponding RPs are seen in (b) and (d), respectively.

indicate transitions between all three types of solution components. The first component with a frequency of 1 is shown by the distance between the diagonal lines. The black section in the center corresponds to no or slow variations of the states in time. For the third component (i.e., the second periodic component), uniform diagonal lines are present and are separated by a length of 0.5, consistent with the frequency of 2.

Figure 12c presents an idealized solution with the following three components: decaying oscillations, small variations with small amplitudes, and growing oscillations. The idealized data resemble the GMM data. The first and third components correspond to a spiral sink and a spiral source, respectively. **Figure 12d** displays the solution's RP with horizontal and vertical lines. Such a RP pattern resembles a combined feature of the RP in **Figure 7b** for a spiral sink and **Figure 10b** for a spiral outward solution.

The RPs for the above idealized data, shown in **Figure 12b** and **d**, are used for a comparison with the RPs of the ERA-interim and GMM data, as shown in **Figure 13a** and **b**, to reveal the features of observed and simulated AEWs in **Figure 11**. **Figure 13a** presents the RP for normalized ERA-interm data at the 850 hPa level. A solid black region is present in the middle, indicating slow variations in states across this region. In areas surrounding the center, the familiar diagonal line structure associated with oscillatory behavior, and possessing some degree of periodicity, can be identified. The data results in a diagonal line structure in RPs that is comparable to **Figure 12b**, consisting of periodic solutions. Based on visual inspection, most of the recurrence points fall into diagonal lines, while a few stand alone. Finding an estimated period for the recurrence of AEWs through the RP is feasible. By calculating the distance between them, we obtain a good measure of the time interval between AEW occurrences. Using this approach, the distance between diagonal lines ranges from 3.5 to 5 days, consistent with the observed AEWs.

Figure 13b provides the RP for normalized data from the 850 hPa level generated by the GMM. The line structure present displays some similarities to **Figure 12d**, with a darker area in the middle and a distinct line pattern surrounding the center. Compared to **Figure 13a**, one major difference is the appearance of horizontal and vertical lines instead of diagonal lines. As suggested by the idealized solution in **Figure 12c** and **d**, this type of structure indicates decaying and growing oscillations, which is also similar to the combined feature in **Figures 7b** and **10b**. The distance between the vertical and horizontal lines can be used to estimate a period for GMM data of approximately 5 days, consistent with observations as well as the RP analysis of the ERA-interim data.

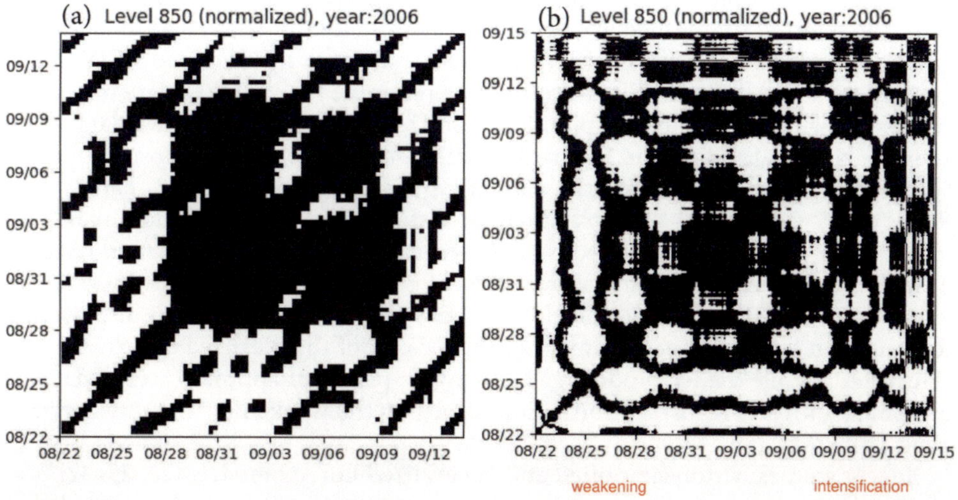

Figure 13.
*Recurrence plots for the normalized wind velocity data at the 850 hPa level for the ERA-interim data (a) and global mesoscale model data (b). Panels (a) and (b) produce the RP analyses that are consistent with those in **Figure 12b** and **d**, respectively.*

4. Concluding remarks

Accurate detection of recurrent, multiple AEWs and their evolution (e.g., intensification) may provide a good indicator for determining the timing and location of tropical cyclone formation and initial intensification. Since large-scale AEWs have better predictability, such an indicator may help extend the lead time of TC formation prediction. To achieve this goal, in this study, we first deployed the recurrence plot method in Python and verified our implementation using several types of idealized solutions and three types of solutions obtained in the classical 3DLM and 5DLM. We then applied the RP method to analyze two, real-world datasets, ERA-interim data and global mesoscale model data, in order to reveal the time-varying amplitudes and frequencies of multiple AEWs over a 30-day period between late August and September 2006. Compared to traditional global analysis methods (e.g., spectral analysis), the RP method is effective in showing temporal growing or decaying oscillations and the transition between these two types of solutions. As a result, the RP analysis of global data not only produces a good estimate of the period for AEWs but also displays the weakening or intensifying trend of AEW intensities, as shown in **Figure 13**.

A summary on the performance of the RP in analyzing various types of solutions is provided below. We first performed the analysis for (1) four basic types of data, including periodic, quasiperiodic, and chaotic solutions as well as Gaussian white noise, (2) idealized decaying/growing oscillations, (3) steady-state solutions using the 3DLM, (4) limit cycle solutions using the 3DLM and 5DLM with a focus on the transition from a spiral solution near a nontrivial critical point to a nonlinear periodic solution that encloses trivial and nontrivial critical points, and (5) idealized solutions that mimic the features of selected AEW data with different periods and different amplitudes. For basic types of solutions, the RP produces a consistent analysis as compared to previous studies. For time-varying oscillations, we discussed how recurrence may lead to horizontal and vertical lines in RPs. The distance between two consecutive horizontal (or vertical) lines gives an estimate of half of a period (i.e., $T/2$). For the fourth and fifth types of datasets, the corresponding RP analyses clearly display local transitions between solution components with various periods or amplitudes.

Based on its promising performance in revealing the features of idealized solutions, the selected RP method was applied to analyze ERA-interim data and GMM data, yielding a RP analysis consistent with the analysis of the idealized AEWs from the fifth dataset. While the ERA-interim data resemble the idealized data in **Figure 12a** with comparable amplitudes but different periods, the GMM data resemble the idealized data in **Figure 12c** with decaying and growing oscillations. The RP analysis of the ERA-interim data in **Figure 13a** produces an estimated period of 3.5–5 days, and a spectral analysis of the same data yields a dominant period of 4.7 days. Both estimated periods are consistent with the observed results. However, only the RP analysis can reveal local variations or transitions. By comparison, the RP pattern of the global model data in **Figure 13b** that displays horizontal and vertical lines is similar to the RP plot of the idealized data in **Figure 12d**. While the RP produces a comparable period of approximately 5 days, it additionally reveals realistic information regarding the weakening and intensifying trend of AEWs.

Recent studies within simplified and generalized Lorenz models [22, 29–32] suggest the importance of detecting oscillatory components for extending the lead time of weather prediction. The analysis with RP is encouraging but largely produces qualitative information. Our future work includes an application of recurrence quantification analysis (RQA) methods (e.g., [33–35]) for quantitatively determining the "recurrence rate" and "determinism" in order to quantitatively measure the recurrence and determinism (or "predictability") of the recurrent solutions.

Acknowledgements

We thank Dr. Y.-L. Wu, Dr. J. Chern, and Ms. S. Faghih-Naini for their scientific discussions and technical help. Thanks are also extended to Dr. B. Bailey, Dr. F. De Sales, and anonymous reviewers for their comments on the original manuscript. We are grateful for the support from the College of Science of San Diego State University.

A. Appendix A: the 3DLM Version 2

The perturbation method was applied to derive the so-called V2 system in order to analyze solutions of the 3DLM, as well as the 5DLM, with initial conditions near the nontrivial critical point. The 5DLM V2 and its non-dissipative V2 were previously documented in [13] and [7], respectively. This appendix simply discusses the 3DLM V2. For the 3DLM, in Eqs. (1)–(3) without the feedback term XY_1, the total field is decomposed into the basic state (X_c, Y_c, Z_c) and the perturbation (X', Y', Z') with respect to the basic state in the form of $X = X_c + X'$ for each state variable, yielding a new set of equations:

$$\frac{dX_c}{d\tau} + \frac{dX'}{d\tau} = -\sigma(X_c + X') + \sigma(Y_c + Y'), \tag{A1}$$

$$\frac{dY_c}{d\tau} + \frac{dY'}{d\tau} = -(X_c Z_c + X_c Z' + Z_c X' + X'Z') + r(X_c + X') - (Y_c + Y'), \tag{A2}$$

$$\frac{dZ_c}{d\tau} + \frac{dZ'}{d\tau} = (X_c Y_c + X_c Y' + Y_c X' + X'Y') - b(Z_c + Z'). \tag{A3}$$

Note that the above basic state solutions are determined using the time-independent, nontrivial critical point solutions, defined as (e.g., [11])

$$Z_c = r - 1, \tag{A4}$$

$$X_c = Y_c = \pm\sqrt{bZ_c}. \tag{A5}$$

Since the basic state solutions are time independent, Eqs. (A1)–(A3) are reduced to become

$$\frac{dX'}{d\tau} = -\sigma(X' - Y'), \tag{A6}$$

$$\frac{dY'}{d\tau} = (r - Z_c)X' - X_c Z' - Y' - \mathbf{FN}(X'Z'), \tag{A7}$$

$$\frac{dZ'}{d\tau} = Y_c X' + X_c Y' - bZ' + \mathbf{FN}(X'Y'), \tag{A8}$$

which describe the time evolution of the perturbations. *FN* is a flag identifying whether the system is fully nonlinear ($FN = 1$) or not ($FN = 0$). The above system

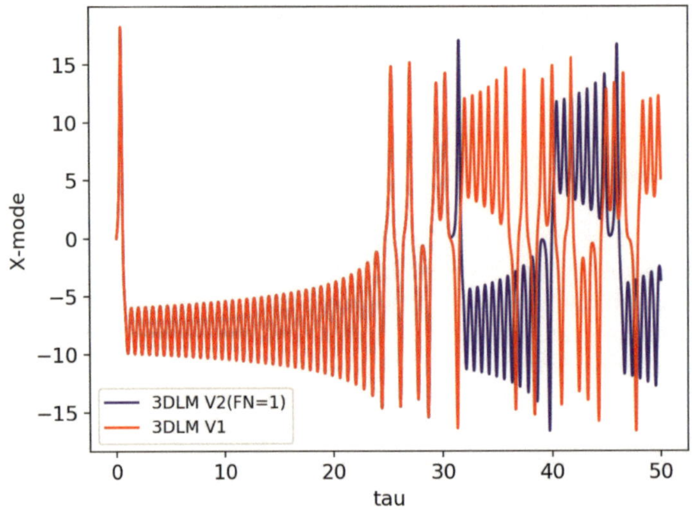

Figure A1.
Time evolution of the X mode solutions of the original 3DLM and 3DLM V2 with $FN = 1$ over the time interval $0 \leq \tau \leq 50$ with a time step $\Delta\tau = 0.001$.

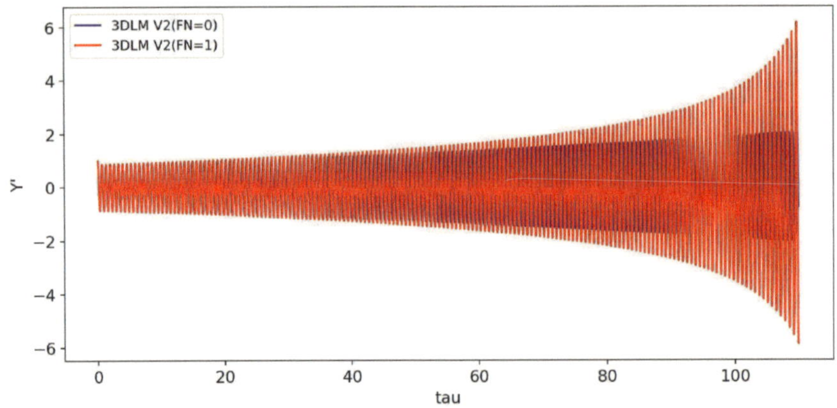

Figure A2.
Time evolution of Y' linear (blue) and nonlinear (red) solutions using the 3DLM V2 with $0 \leq \tau \leq 110$ and $\Delta\tau = 0.001$.

with $FN = 0$ or $FN = 1$ is referred to as a Version 2 system, denoted by V2. In the V2 system, there are no nonlinear terms that involve the product of two basic state variables. The system only allows interactions between one basic state variable and one perturbation (i.e., X_cY') or between two perturbation variables (e.g., $X'Y'$). By setting $FN = 0$, the system is linear with respect to the perturbation. When the initial perturbations are small, nonlinear terms with two perturbation variables are smaller. Thus, such a linear system is capable of depicting the initial time evolution of the full 3DLM when the perturbation is small. In comparison to the full 3DLM, the nonlinear V2 model ($FN = 1$) produces the same solutions, even at the onset of chaotic irregular oscillations, as shown in **Figure A1**. The time evolutions of the Y' solutions for both $FN = 0$ and $FN = 1$ are shown in **Figure A2**. The solutions are initially the same and begin to diverge as time further progresses, indicating the impact of nonlinearity. An eigenvalue analysis of the 3DLM V2 is provided in Section 2.1.3 of Reyes [24].

Author details

Tiffany Reyes and Bo-Wen Shen*
Department of Mathematics and Statistics, San Diego State University, San Diego, CA, USA

*Address all correspondence to: bshen@sdsu.edu

References

[1] Shen B-W, Atlas R, Reale O, Lin S-J, Chern J-D, Chang J, et al. Hurricane forecasts with a global mesoscale-resolving model: Preliminary results with hurricane Katrina (2005). Geophysical Research Letters. 2006;**33**: L13813. DOI: 10.1029/2006GL026143

[2] Shen B-W, Tao W-K, Wu M-L. African easterly waves in 30-day high-resolution global simulations: A case study during the 2006 NAMMA period. Geophysical Research Letters. 2010;**37**: L18803. DOI: 10.1029/2010GL044355

[3] Shen B-W, Cheung S, Li J-LF, Wu Y-L, Shen SS. Multiscale processes of Hurricane Sandy (2012) as revealed by the parallel ensemble empirical mode decomposition and advanced visualization technology. Advances in Data Science and Adaptive Analysis. 2016;**08**:1650005. DOI: 10.1142/S2424922X16500054

[4] Shen B-W, Cheung S, Wu Y, Li F, Kao D. Parallel implementation of the ensemble empirical mode decomposition (PEEMD) and its application for earth science data analysis. Computing in Science & Engineering. 2017;**19**(5):49-57. DOI: 10.1109/MCSE.2017.3421555

[5] Wu Y-L, Shen B-W. An evaluation of the parallel ensemble empirical mode decomposition method in revealing the role of downscaling processes associated with African easterly waves in tropical cyclone genesis. Journal of Atmospheric and Oceanic Technology. 2016;**33**: 1611-1628. DOI: 10.1175/JTECH-D-15-0257.1

[6] Marwan N, Webber CL Jr. Recurrence Quantification Analysis: Theory and Best Practices. Switzerland: Springer International Publishing; 2015. p. 421

[7] Faghih-Naini S, Shen B-W. Quasi-periodic in the five-dimensional non-dissipative Lorenz model: The role of the extended nonlinear feedback loop. International Journal of Bifurcation and Chaos. 2018;**28**(6): 1850072. DOI: 10.1142/S0218127418500724

[8] Thompson JMT, Stewart HB. Nonlinear Dynamics and Chaos. 2nd ed. Chichester, United Kingdom: John Wiley & Sons, Ltd.; 2002. p. 437

[9] Shen B-W, Atlas R, Chern J-D, Reale O, Lin S-J, Lee T, et al. The 0.125 degree finite-volume general circulation model on the NASA Columbia supercomputer: Preliminary simulations of mesoscale vortices. Geophysical Research Letters. 2006;**33**:L05801. DOI: 10.1029/2005GL024594

[10] Zipser EJ et al. The Saharan air layer and the fate of African easterly waves—NASA AMMA field study of tropical cyclogenesis. Bulletin of the American Meteorological Society. 2009;**90**: 1137-1156

[11] Lorenz E. Deterministic nonperiodic flow. Journal of the Atmospheric Sciences. 1963;**20**:130-141

[12] Lorenz EN. The Essence of Chaos. Seattle: University of Washington Press; 1993. p. 227

[13] Shen B-W. Nonlinear feedback in a five-dimensional Lorenz model. Journal of the Atmospheric Sciences. 2014;**71**: 1701-1723. DOI: 10.1175/JAS-D-13-0223.1

[14] Lin SJ. A vertically Lagrangian finite volume dynamical core for global models. Monthly Weather Review. 2004; **132**:2293-2307. DOI: 10.1175/1520-0493

[15] Moon S, Han B-S, Park J, Seo JM, Baik J-J. Periodicity and chaos of high-order Lorenz systems. International Journal of Bifurcation and Chaos. 2017;

27(11):1750176. DOI: 10.1142/S021812741 7501760

[16] Felicio CC, Rech PC. On the dynamics of five- and six-dimensional Lorenz models. Journal of Physics Communications. 2018;**2**:025028

[17] Shen B-W. Nonlinear feedback in a six-dimensional Lorenz model. Impact of an additional heating term. Nonlinear Processes in Geophysics. 2015;**22**: 749-764. DOI: 10.5194/npg-22-749-2015

[18] Shen B-W. Hierarchical scale dependence associated with the extension of the nonlinear feedback loop in a seven-dimensional Lorenz model. Nonlinear Processes in Geophysics. 2016;**23**:189-203. DOI: 10.5194/npg-23-189-2016

[19] Shen B-W. On an extension of the nonlinear feedback loop in a nine-dimensional Lorenz model. Chaotic Modeling and Simulation (CMSIM). 2017;**2**:147157

[20] Shen BW, Reyes T, Faghih-Naini S. Coexistence of chaotic and non-chaotic orbits in a new nine-dimensional Lorenz model. In: Skiadas CH, Lubashevsky I, editors. The 11th Chaos International Conference; Springer Proceedings in Complexity; Cham: Springer; 2019. DOI: 10.1007/978-3-030-15297-0_22

[21] Reyes TAL, Shen B-W. A recurrence analysis of chaotic and non-chaotic solutions within a generalized nine-dimensional Lorenz model. Chaos, Solitons & Fractals. 2019:**125**(2019): 1-12. DOI: 10.1016/j.chaos. 2019.05.003

[22] Shen B-W. Aggregated negative feedback in a generalized Lorenz model. International Journal of Bifurcation and Chaos. 2019;**29**(3):1950037. DOI: 10.1142/S0218127419500378

[23] Marwan N, Romano MC, Thiel M, Kurths J. Recurrence plots for the analysis of complex systems. Physics Reports. 2007;**438**:237-329

[24] Reyes TAL. Applying recurrence quantification analysis methods for the analysis of global reanalysis and model data to reveal the local oscillations of multiple African easterly waves during 2006 [Master thesis]. San Diego State University; 2018. p. 64

[25] Webber CL Jr. Recurrence quantification of fractal structures. Front Physiotherapy. 2012;**3**:382. DOI: 10.3389/fphys.2012.00382

[26] Jordan DW, Smith P. Nonlinear Ordinary Differential Equations. An Introduction for Scientists and Engineers. 4th ed. New York: Oxford University Press; 2007. p. 560

[27] Nagel RK, Saff E, Snider A. Fundamentals of Differential Equations. 7th ed. New York: Pearson; 2008. p. 912

[28] Shen B-W, Tao W-K, Lau W, Atlas R. Predicting tropical cyclogenesis with a global mesoscale model: Hierarchical multiscale interactions during the formation of tropical cyclone Nargis (2008). Journal of Geophysical Research. 2010;**115**:D14102. DOI: 10.1029/2009JD013140

[29] Shen B-W. On the predictability of 30-day global mesoscale simulations of multiple african easterly waves during summer 2006: A view with a generalized Lorenz model. Geosciences 2019b;**9**(7):281. DOI: 10.3390/geosciences9070281

[30] Shen B-W. Homoclinic orbits and solitary waves within the non-dissipative Lorenz model and KdV equation. International Journal of Bifurcation and Chaos. DOI: 10.1142/S0218127420502570. (in press)

[31] Shen B-W, Pielke RA Sr, Zeng X, Baik J-J, Faghih-Naini S, Cui J, Atlas R. Is weather chaotic? Coexistence of chaos and order within a generalized Lorenz

model. Bulletin of American Meteorological Society. 2020: 1-28. DOI: 10.1175/BAMS-D-19-0165.1

[32] Shen B-W, Pielke RA Sr, Zeng X, Baik J-J, Faghih-Naini S, Cui J, Atlas R, Reyes TA. Is Weather chaotic? Coexisting chaotic and non-chaotic attractors within Lorenz models. The 13th Chaos International Conference (CHAOS2020); 9-12 June 2020. (virtual conference)

[33] Webber CL Jr, Zbilut JP. Dynamical assessment of physiological systems and states using recurrence plot strategies. Journal of Applied Physiology. 1994; **76**(2):965-973

[34] Zbilut JP, Webber CL Jr. Embeddings and delays as derived from quantification of recurrence plots. Physics Letters A. 1992;**171**(34):199-203. DOI: 10.1016/0375-9601(92)90426-M

[35] Zbilut JP, Webber CL Jr. Recurrence quantification analysis: Introduction and historical context. International Journal of Bifurcation and Chaos. 2007; **17**(10):3477-3481. DOI: 10.1142/S0218127407019238

Chapter 4

Response of the Coastal Ocean to Tropical Cyclones

Zhiyuan Wu and Mack Conde

Abstract

The Northwest Pacific and the South China Sea region are the birthplaces of most monsoon disturbances and tropical cyclones and are an important channel for the generation and transmission of water vapor. The Northwest Pacific plays a major role in regulating interdecadal and long-term changes in climate. China experiences the largest number of typhoon landfalls and the most destructive power affected by typhoons in the world. The hidden dangers of typhoon disasters are accelerating with the acceleration of urbanization, the rapid development of economic construction and global warming. The coastal cities are the most dynamic and affluent areas of China's economic development. They are the strong magnetic field that attracts international capital in China, and are also the most densely populated areas and important port groups in China. Although these regions are highly developed, they are vulnerable to disasters. When typhoons hit, the economic losses and casualties caused by gale, heavy rain and storm surges were particularly serious. This chapter reviews the response of coastal ocean to tropical cyclones, included sea surface temperature, sea surface salinity, storm surge simulation and extreme rainfall under the influence of tropical cyclones.

Keywords: tropical cyclones, typhoons, coastal ocean dynamics, sea surface temperature, sea salinity, extreme rainfall; air-sea interaction

1. Introduction

Tropical cyclones are some of the most destructive natural disasters, which often bring huge losses to people's life and property. The Northwest Pacific and the South China Sea regions are the birthplaces of most monsoons and typhoons and are important channels for the generation and transmission of water vapor [1–3]. The influence of a typhoon on a region is often not only a heavy wind disaster. At the same time, the heavy rain, extreme waves, storm surges, and coastal inundation that are produced will also have a huge impact on the region, which will result in the formation of a typhoon disaster chain [4–6].

There are more than 20 typhoons in the Pacific Northwest each year, which is the region with the most frequent typhoon activity in the world. China, which has a long coastline on the west coast of the Pacific Ocean, is the country with the most frequent typhoon attacks in the world, with an average of 9.3 per year, resulting in a direct economy every year. The losses exceeded 100 billion yuan and the number of casualties reached thousands. Therefore, a comprehensive understanding and in-depth study of the typhoon process, especially the improvement of typhoon monitoring and early warning capabilities, is an inevitable requirement for the disaster reduction work of our national defense platform.

In the past few decades, from the perspective of atmospheric science, the research on the mechanisms of typhoon development, numerical simulation and forecasting has made great progress. However, as a strong atmospheric process, typhoons have violent disturbances, the cumulative effects of many typhoons will also have a significant impact on the ocean's thermo-salt structure and ocean circulation and global ocean heat transport. These effects will counteract typhoons, affect not only the intensity and path of specific typhoons, but also the global the low-frequency variation characteristics of the typhoon. But so far, the lack of on-site observation data during the typhoon has made the study of multi-scale response and feedback mechanism of typhoon not deep enough. People cannot simulate the interaction process between the ocean and typhoon well. The reliable initial ocean field required for typhoon forecasting has greatly limited the further improvement of typhoon research and forecasting capabilities.

The typhoon is a devastating natural disaster that has long been a focus of attention in the field of atmospheric and oceanic research [7, 8]. With the rapid development of computers, the numerical simulation of typhoons is becoming increasingly developed, and the model resolution is getting higher and higher [9]. The Pacific Northwest is the most concentrated area of global tropical cyclones (also known as typhoons in the Pacific Northwest). China is located on the west coast of the Pacific Ocean, with a long coastline and a special geographical position on the southeast coast. It has been hit by typhoons frequently, with an average annual rate of 9.3, ranking first in the world. The typhoon is one of the most serious natural disasters in China [10–12]. The annual direct economic losses caused by the typhoon are nearly 100 billion yuan, and the number of casualties is thousands. On the one hand, the strong winds and heavy rains that landed in the typhoon brought huge meteorological disasters to the vast areas of China, posing a huge threat to the people's lives, property and production activities. On the other hand, huge waves and storm surges caused by typhoons have also caused serious marine disasters, which have caused major safety hazards and economic losses to offshore operations and transportation, coastal protection projects, marine fisheries and marine aquaculture. Coastal areas are one of China's most economically developed regions, which are vulnerable to the effects of marine disasters [13–16].

Therefore, studying the movement mechanism of typhoon, accurately forecasting the influence of typhoon and reducing storm surge disasters have important social value for the protection of national economic development and people's lives and property safety.

From the perspective of practical application, improving the monitoring and forecasting ability of typhoon is the fundamental goal of typhoon research. Due to the multi-scale characteristics of the interaction between ocean and typhoon, the ocean data assimilation for the typhoon process should also be multi-scale. Due to extreme sea conditions under typhoon conditions Harsh, satellite remote sensing data has become an important data source for assimilation. However, remote sensing can only provide sea surface information. At present, people usually use the projection mapping method and multiple dynamic constraints to map surface information to the ocean subsurface and assimilate. Some assimilation methods still lack universality. How to establish a sea surface data assimilation method that considers more dynamic constraints and is more suitable for typhoon conditions is an important part of the future sea-air coupled data assimilation research. For the actual operational forecast of the typhoon, International or domestic still rely mainly on numerical weather patterns. After long-term exploration and improvement, the main forecasting modes are for atmospheric processes (such as atmospheric boundary layer physical properties, cloud physical processes, atmospheric turbulent energy calculations, cumulus convective parameterization schemes, etc.). There has been considerable progress in simulation and forecasting capabilities. The understanding of ocean feedback is insufficient. The current typhoon numerical (weather) forecasting model still has significant errors, especially for typhoon intensity and wind and rain distribution forecast. The United States is the first to develop a sea-air coupled hurricane (the Atlantic called hurricane, the Pacific Ocean). After years of operational operation, the improved air-sea coupled model has improved both the intensity of the hurricane and the path prediction compared with the traditional atmospheric model.

In summary, the interaction between the ocean and typhoon has obvious scientific significance and important practical value. At various time and space scales, people's understanding of the mechanism of the ocean response to typhoon and modulation is obviously insufficient. At present, air-sea coupling The model's ability to simulate, assimilate and forecast the typhoon process is still very limited, which has become a bottleneck problem to further improve the typhoon forecasting capability.

2. Sea surface temperature response to tropical cyclones

With an increase in sea surface temperature (SST), the total number of tropical cyclones in the North Pacific, Indian Ocean, and southwest Pacific Ocean decreases, and the cyclone development period shortens, but the number and proportion of tropical cyclones reaching super typhoon intensity increases greatly [17].

In the study of air-sea interaction, the response of the upper ocean to typhoons is a hot topic [18]. Typhoon transit can cause ocean mixing and upwelling, and sea surface cooling is the main feature [19]. The cooling caused by a typhoon is mainly related to the intensity, propagation speed of the typhoon, and the ocean condition before typhoon arrival, such as the location of cold vortices, the thermodynamic structure of the upper ocean, the position of the 26°C isotherm, etc.

Cyclonic wind stress results in the upwelling of sea water in the center of the path, the decrease of sea surface temperature, and the heat transfer from the surface to the atmosphere. Strong winds cause turbulent mixing of the ocean, entraining cold water from the lower layer into the mixing layer, resulting in cooling of the upper sea water and deepening of the mixing layer [10–21]. Inclusive mixing disturbances cause 85% of irreversible ocean heat to enter the atmosphere; direct air-sea interaction plays a minor role in surface cooling. Mixed layer plays an important role in sea surface cooling [22–24]. After typhoon transit, the ocean response is mostly the internal nonlocal baroclinic process caused by wind stress. In baroclinic driving stage, the flow of mixing layer 1 m/s is induced by vertical mixing, and the flow of near-inertial oscillation frequency wave into thermocline, which lasts for 5–10 days. The barotropic driving process usually results in geostrophic currents and associated sea surface height changes.

Using the observed data to study the ocean response to typhoons is a common research method. However, due to the severe weather conditions during the transit of tropical cyclones, it is very difficult to obtain fixed-point observation data. There is a strong mass transport, energy exchange, and interaction between the atmosphere and the ocean during a typhoon process [25, 26] (**Figure 1**).

The response mechanism of the ocean to typhoon can be considered from two levels. First, the typhoon-driven mesoscale three-dimensional ocean circulation will have a significant impact on the local dynamics and thermal processes. The resulting near-inertial internal waves and vortices can input a large amount of mechanical

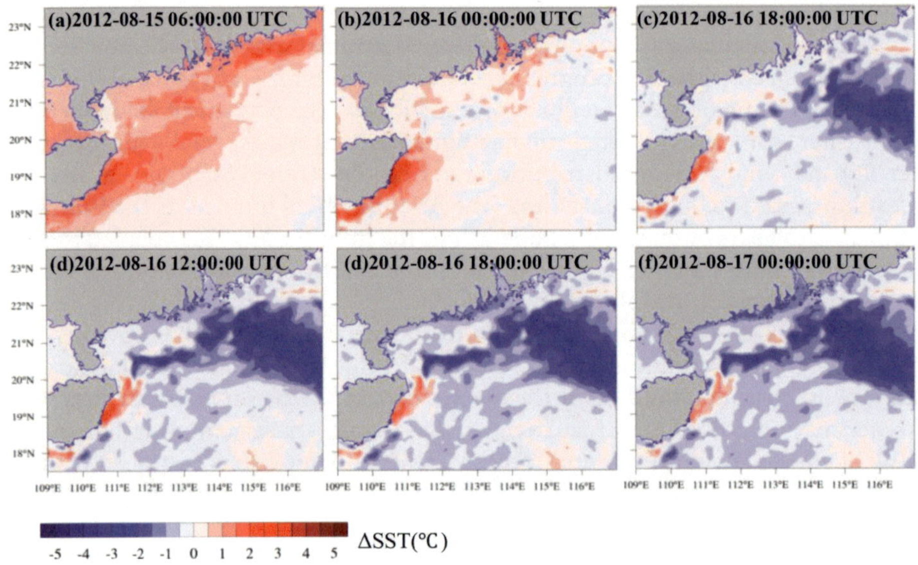

Figure 1.
The spatial distribution of change in sea surface temperature (ΔSST) in the northern area of the South China Sea under the influence of typhoon Kai-tak (2012).

energy into the ocean, thus significantly enhancing the local The ocean mixes and changes the warm salt structure of the upper ocean. Second, in the interior of the ocean, the energy input into the ocean by typhoons in the form of near-inertial internal waves travels along the oceanic thermocline to distant places, such as the entire tropical Pacific. During the propagation process, they interact nonlinearly with the original internal waves and near-inertial oscillations inside the ocean, which affects the ocean basin scale and even the global energy distribution, and leads to an increase in ocean mixing rate in some specific regions. The modulation of the typhoon by the ocean can also be considered from two scales. On the weather scale, the ocean plays a very important role in the movement and action of typhoons.

The maturity stage is mainly characterized by negative feedback that reduces the sea surface temperature. However, when the upper ocean warm water is thicker, the typhoon transit will not cause obvious sea temperature anomaly, and the lack of negative ocean feedback can cause the typhoon to strengthen. The interaction between the ocean mesoscale process and the typhoon is currently a focus of typhoon research. Usually, the warm vortex can quickly strengthen the typhoon, and the cold vortex can quickly weaken the typhoon. At the climate scale, global warming and interannual and interdecadal variations of climate can cause changes in ocean circulation and thermal conditions, resulting in low-frequency modulation of the intensity and frequency of typhoons.

3. Sea surface salinity response to tropical cyclones

The typhoon is one of the most serious natural disasters that affects the coastal ocean environment in China [27, 28], especially in the eastern and southern estuaries, such as the Yangtze River Estuary [29] and the Pearl River Estuary [30–33]. During a typhoon, the coupling of various dynamic factors, such as wind, waves, storm surges, and river runoff, greatly enhances the mass and energy exchange of various interfaces in the ocean and is accompanied by heavy rain and storm runoff on the surface [34–37]. Scouring can transport a large amount of minerals from the land to an estuary offshore, causing sudden changes in the water quality of the estuary, which may have an important impact on the marine ecological environment [38–40].

On the one hand, typhoon transit strengthens the mixing process of offshore water [41–43]. On the other hand, the heavy rainfall brought by a typhoon rapidly increases river runoff into the sea, and a large amount of land-based materials are washed away and brought into the estuary offshore area [44–47]. These changes due to the influence of a typhoon significantly affect the physical, chemical and biological processes of estuarine offshore waters, which in turn have an impact on the structure and function of the ecosystem [48–50]. Studying the changes of the estuarine nearshore environment under the influence of a typhoon and its ecological effects are of great importance for further understanding the evolution process of ecosystems in this region on a long-term scale [8, 51].

Field observations show that the salinity of the surface water of an estuary usually shows a sharp change during a typhoon and the resulting rain, which gradually rises after entering the recovery period [8, 52–54]. During typhoon crossing, the disturbance caused by strong winds strengthens the mixing process of the estuary and its adjacent waters. However, this process has a passing impact on the water environment, and the runoff diluting water expansion and the external seawater intrusion play a greater role in changing the water environment after a typhoon. Among these, the strengthening of a typhoon after the expansion of fresh water greatly affects the upper water, the upper salinity decreases after the typhoon, and the nutrient salt concentration increases significantly. External seawater intrusion substantially changes the bottom water environment. The salinity of the bottom

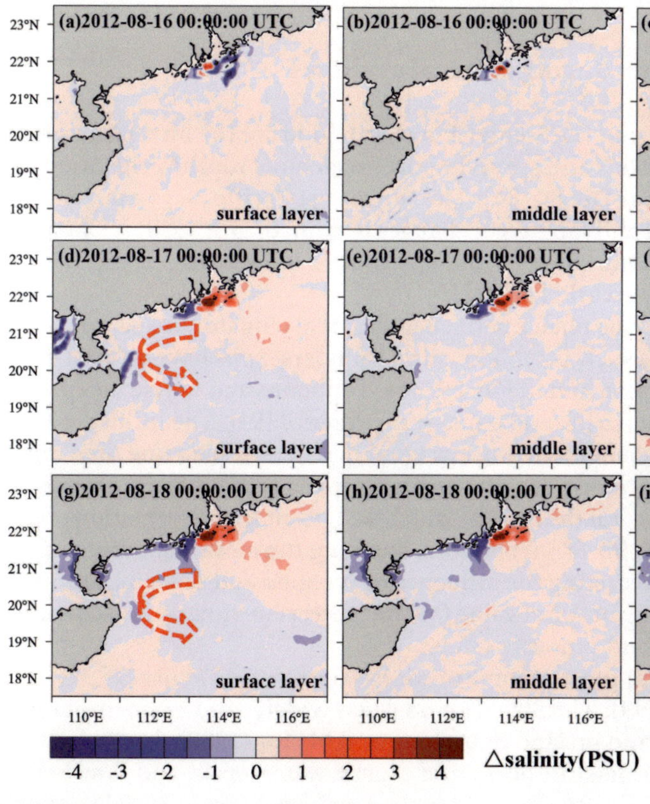

Figure 2.
Changes in stratifications salinity influenced by typhoon Kai-tak based on the fully coupled WRF-SWAN-ROMS model (beginning on 2012-08-15 00:00:00 UTC).

layer increases after a typhoon, and the nutrient concentration of nitrogen and silicon decreases.

Typhoons or tropical cyclones are strong wind events in the climate system and are a strong form of air-sea interaction. The strong vertical mixing and wind field generated by a typhoon has a major impact on the upper ocean dynamics and ecosystem [55]. Due to typhoons, there is a decrease in sea level, a decrease in sea surface temperature, an increase in phytoplankton blooms and a decrease in primary productivity, which also affect marine fisheries [56–58]. Typhoons mainly affect the marine ecological environment through two physical mechanisms: (1) after a typhoon, a cold vortex is formed, causing seawater to upwell and the lower layer of cold nutrient water is transported to the upper layer [59, 60]; and (2) the typhoon intensifies the vertical mixing of the upper ocean by a strong wind process [61–64].

At present, most research on the sea surface salinity (SSS) response to typhoons is limited to the estuary area. According to the physical and biochemical environmental conditions of the estuary, SSS may show an upward or downward trend after typhoon transit [2, 22, 65–75]. However, studies on marine ecological factors, especially SSS and the response to typhoon transit, are limited and have not been discussed in detail [76–79]. The South China Sea (CSC) is the largest marginal sea in the Pacific Northwest, and is also a frequent typhoon zone, but it is difficult to obtain measured data during typhoons.

Due to the harsh meteorological conditions during typhoon transit, the use of on-site observation methods in an estuary to study the changes in the marine environment before and after a typhoon is very limited. The numerical simulation method is an effective way to study the distribution characteristics of fresh and salt water in an estuary under the influence of a typhoon (**Figure 2**).

4. Storm surge modeling during tropical cyclones

The Northwest Pacific and the South China Sea region are the birthplaces of most monsoons and typhoons and are an important channel for the generation and transmission of water vapor [18, 80–84]. The Northwest Pacific plays a major role in regulating interdecadal and long-term changes in climate [46, 85, 86]. China is the region with the largest number of typhoons and the most destructive power affected by typhoons in the world [87, 88].

Compared with large-scale phenomena such as global climate change, small- and medium-scale phenomena such as typhoons and thunderstorms have an even greater impact on people's production and life [6, 89, 90]. Typhoons and hurricanes present some of the greatest threats to life and damage to property [91]. The influence of a typhoon on a region is often not only a heavy wind disaster. At the same time, the heavy rain, extreme waves, storm surges and beach erosion [24] that are produced will also have a huge impact on the region, which will result in the formation of a typhoon disaster chain [19, 92–94]. Therefore, studying the movement mechanism of typhoon, accurately forecasting the influence of typhoon and reducing storm surge disasters have important social value for the protection of national economic development and human and property safety.

Tropical cyclones (TCs) present some of the greatest threats to life [25, 95–98] and damage to property [99]. The SLOSH model was widely used in storm surge simulation in seas, lakes, and on land. Blumberg and Mellor (1987) developed the POM model to simulate large-scale ocean and coastal water levels, and flow field changes. Many ocean models have been developed and used for the simulation of storm surges, such as the ECOM model, ROMS model, CH3D-IMS model, CEST model, SELFE model, Delft3D model, ADCIRC model and FVCOM model. They have achieved very good results and laid the foundation for understanding the dynamic mechanism of storm surges. The development of a coupled atmospheric and ocean model had significant effects on improving the accuracy of numerical prediction. The establishment of a coupled atmosphere and ocean model is an effective method to solve this problem (**Figure 3**).

The typhoon numerical model is the focus of typhoon research and the key to typhoon forecasting. The modern model has a certain forecasting ability for the typhoon path, but the forecast of typhoon intensity is still a recognized problem in the international meteorological community. The reason is that, besides the understanding and simulation of the atmospheric environment and the structure of the typhoon itself is not accurate enough, it is also one of the important reasons for the lack of understanding of the complexity and feedback of related ocean dynamics and thermal processes. When the typhoon transits, it exerts a great shearing force on the sea surface. The related wave breaking and the interaction between the wind field and the Stokes drifting can generate a large amount of turbulent kinetic energy, which produces a wave below the sea surface. The turbulent enhancement zone enhances the rate of turbulence dissipation in the upper ocean. Therefore, the establishment of a relatively complete marine hybrid scheme is an important way to improve the maritime-coupled typhoon model. In addition, improving the sea surface flux parameterization scheme under strong wind conditions is also an urgent need to improve the model prediction capability. With the rapid increase of computing power and technology, the air-sea coupled typhoon model has broken through the limitations of the early axisymmetric typhoon model and the mixed-layer ocean model, and replaced it with a complete fully coupled ocean

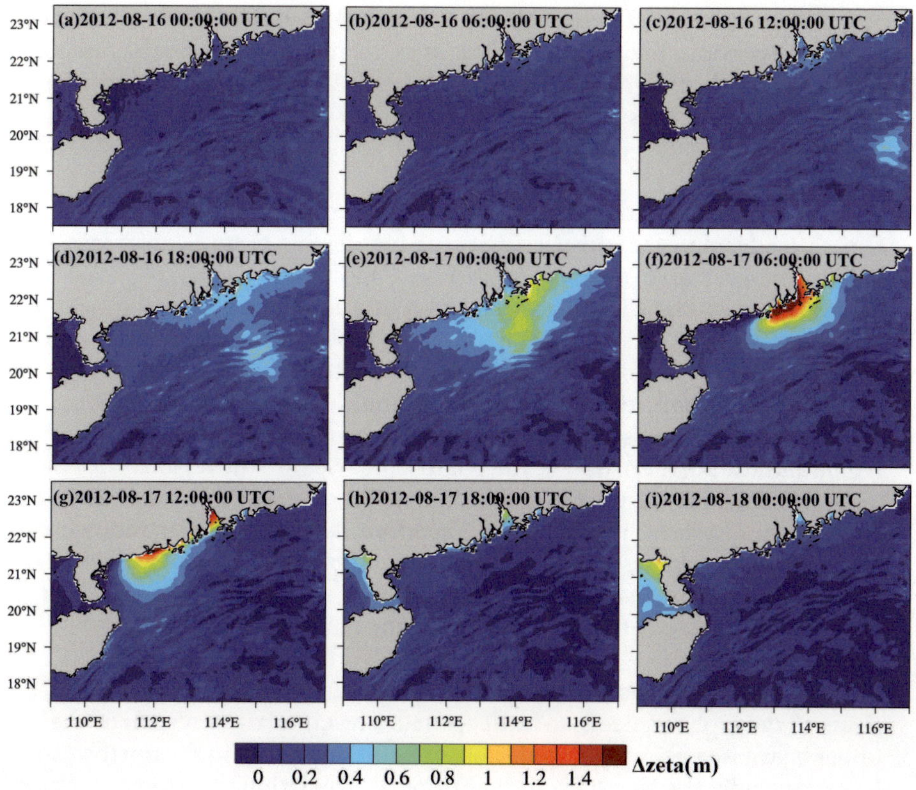

Figure 3.
Spatial distribution of storm surge level influenced by typhoon Kai-tak (start at 2012-08-15 00:00:00 UTC).

and atmosphere model. At present, the world's 1/32 to 1/900 degree resolution ocean model is being developed, which will provide strong support for the study of small-scale processes in the ocean and the multi-scale interaction between ocean and typhoon.

5. Future plans

The interaction between the ocean and the typhoon is a major scientific issue with significant scientific significance and important practical value. In recent years, with the support of national major scientific research projects, China has comprehensively utilized on-site observations from the perspective of air-sea interaction. Research methods such as theoretical analysis, data assimilation, and model prediction systematically study the response and modulation mechanism of the upper ocean to the typhoon, the interaction between the ocean and atmospheric observation system for the typhoon, the ocean mesoscale process and the typhoon, and the ocean to the typhoon. A series of innovations have been achieved in low-frequency response and modulation, physical mechanisms and parameterization of typhoons affecting the upper oceans, ocean multi-source data assimilation and parameter estimation during the typhoon, and ocean-air coupled prediction technology and applications in typhoons and marine environments. These research results will provide a solid theoretical foundation and technical support for further improving the forecast level of typhoon business in China, and make substantial contributions to the major national needs of disaster prevention and reduction.

However, it must also be recognized that China is still very lacking in the research field of interaction between ocean and typhoon, and there is still a big gap with the international advanced level. Compared with the national demand for disaster prevention, there are still obvious deficiencies. Based on the research results, we believe that the major scientific problems and major challenges in the interaction between oceans and typhoons are mainly reflected in the following points.

a. On-site observations are still very scarce. As described in this paper, China has already made important practices in ocean monitoring of typhoon processes and has obtained valuable on-site observations. Especially in the field of sea-air coordinated observation, China has launched A useful attempt. After the technology and security conditions are more mature, the typhoon observations coordinated by the sea-air will provide the necessary information for deepening the typhoon research. In addition, due to the harsh sea conditions during the typhoon, the long-term monitoring system for the typhoon process is still missing. The Pacific region and the northern part of the South China Sea are the regions with the highest typhoon in the world, and are almost the only way for typhoons that cause major disasters in our country. Therefore, long-term observation networks are built and maintained in the region (for example, the cross buoy/potential system) An array of observations for the basic structure is an effective means of enhancing ocean and atmospheric monitoring during the typhoon.

b. The response mechanism of the multi-scale circulation system of the upper ocean to the typhoon needs to be deepened. The circulation system of the upper ocean is very complicated. The typhoon prevailing in the northwestern Pacific includes the North Pacific subtropical circulation and tropical circulation driven by the trade wind, by buoyancy flux. The shallow transfected circulation of the North Pacific, the monsoon-driven circulation of the South China

Sea, and the small-scale circulation and vortex superimposed on these large-scale circulations. Typhoons can not only affect and even drive small- and medium-scale ocean circulation and vortex on the weather scale. The rotation can also affect the large-scale ocean circulation of the climatic state by changing the thermal salt structure of the upper ocean. Therefore, the response of the multi-scale circulation system of the upper ocean to the typhoon includes various dynamic processes, thermal processes, and nonlinearities between them. Interactions, these are major challenges in the study of the interaction between ocean and typhoon. Reveal the propagation, transfer and dissipation mechanisms of near-inertial energy input into the ocean by typhoons, and understand the mesoscale processes such as ocean vortex and internal waves during typhoon transit. Response characteristics and excitation mechanism to determine the "heat pump" and "cold suction" of the typhoon. The different effects on ocean stratification are the core of solving these problems.

c. A quantitative study on the modulation of typhoon intensity by the dynamic and thermal structures of the upper ocean. The dynamic and thermal structure of the upper ocean determines the magnitude of sensible heat and latent heat flux at the air-sea interface during typhoon transit. The maintenance and development of typhoons, especially the changes in typhoon intensity, depending on the energy and water vapor provided by these fluxes. Therefore, the dynamic and thermal structures of the upper oceans can play an important role in modulating the intensity of typhoons. The path and intensity are closely related, but since the typhoon intensity is directly affected by the energy provided by the ocean and is the weak link of the current typhoon forecast, we should pay special attention to the modulation of the typhoon intensity by the ocean. If the marine environment does not change, this modulation can be easily estimated from the upper maritime structure of the climatic state. But the problem is that the dynamic and thermal structures of the upper ocean are constantly changing at various spatial and temporal scales. Understand the feedback mechanism of the maritime mesoscale process on the typhoon on the weather scale, reveal the climate. The low-frequency variation of the upper ocean circulation and heat content under changing background should be solved by this question. The key to the question.

In short, based on the existing research foundation and experience, we suggest that in the future research on the interaction between ocean and typhoon. On the basis of the mechanism, the typhoon intensity and the forecasting ability of the marine environment are improved, and the predictability of typhoon low-frequency variability is evaluated, making China one of the world's leading researchers in the interaction between ocean and typhoon.

6. Summary

Observations over the past few decades have shown that the frequency, intensity, and duration of tropical cyclones vary over the interannual, interdecadal, and even longer timescales. Global warming caused by human activities and low-frequency natural oscillations in the Earth's climate system may have an impact on typhoons, but the relative importance of the two is still controversial. Whatever the case, the role of the ocean is unquestionable. Because on a long-term scale, the memory of the climate system is mainly stored in the ocean, any low-frequency variation must be related to the ocean. Previous studies on the low-frequency modulation of tropical cyclones in the ocean have focused on the correlation analysis between tropical sea surface temperature and typhoon parameters, but such analysis has its limitations. For example, the variation of the total power consumption of the Atlantic tropical cyclone has a good correlation with the variation of the sea surface temperature. If this empirical relationship is brought into the climate model, the total power consumption of the Atlantic tropical cyclone will increase by 3 times by the end of the 21st century. However, if a similar empirical relationship is established by subtracting the global tropical average from the tropical Atlantic sea surface temperature variation, the total tropical Atlantic cyclone power consumption predicted by the climate model remains essentially unchanged. This shows that the Atlantic tropical cyclone has been mainly modulated by natural low-frequency oscillations for the past 30 years.

In addition to high-resolution models, advanced data assimilation techniques are also essential to improve the simulation and forecasting capabilities of the typhoon model. Data assimilation can assimilate data from different sources, different time and space, and different elements into the dynamic model, and obtain an analysis field that is more detailed than the observation data and more realistic than the model results. For the assimilation of ocean data in the typhoon process, the most important problem is how to achieve multi-scale, multi-variable assimilation, extract the information reflecting the multi-scale interaction between ocean and typhoon in the observation system, and ensure the consistency of the model state field correction; The determination of the dependent background field error covariance matrix is also a problem.

In summary, the response and modulation mechanism of the ocean to typhoons is an international frontier proposition for marine and atmospheric science research. It is extremely challenging in terms of theoretical methods, observation techniques, model development and data assimilation. Taking this as an entry point, it is expected to achieve breakthrough basic research results, develop and improve marine science theories, and promote the interdisciplinary and common development of marine and atmospheric sciences while meeting the major needs of the country.

Acknowledgements

The study was supported by the National Natural Science Foundation of China (Grant Nos. 51809023, 51839002, and 51879015). The partial support also comes from the Open Research Foundation of Key Laboratory of the Pearl River Estuarine Dynamics and Associated Process Regulation, Ministry of Water Resources ([2018] KJ03), and the Research Foundation of Education Bureau of Hunan Province, China (Grant No. 19C0092).

Author details

Zhiyuan Wu[1,2,3*] and Mack Conde[4]

1 School of Hydraulic Engineering, Changsha University of Science and Technology, Changsha, China

2 Key Laboratory of Water-Sediment Sciences and Water Disaster Prevention of Hunan Province, Changsha, China

3 School for Marine Science and Technology, University of Massachusetts Dartmouth, New Bedford, MA, USA

4 School of Marine Science and Ocean Engineering, University of New Hampshire, Durham, NH, USA

*Address all correspondence to: zwu@csust.edu.cn

References

[1] Wu Z, Jiang C, Deng B, et al. Sensitivity of WRF simulated typhoon track and intensity over the South China Sea to horizontal and vertical resolutions. Acta Oceanologica Sinica. 2019;**38**:74-83

[2] Chia HH, Ropelewski CF. The interannual variability in the genesis location of tropical cyclones in the Northwest Pacific. Journal of Climate. 2002;**15**:2934-2944

[3] Wang G, Su J, Ding Y. Tropical cyclone genesis over the South China Sea. Journal of Marine Systems. 2007;**68**:318-326

[4] Yasuda T, Nakajo S, Kim SY. Evaluation of future storm surge risk in East Asia based on state-of-the-art climate change projection. Coastal Engineering. 2014;**83**:65-71

[5] Sun J, Wang G, Zuo J. Role of surface warming in the northward shift of tropical cyclone tracks over the South China Sea in November. Acta Oceanologica Sinica. 2017;**36**:67-72

[6] Lee TL. Neural network prediction of a storm surge. Ocean Engineering. 2006;**33**:483-494

[7] Almar R, Marchesiello P, Almeida LP, Thuan DH, Tanaka H, Viet NT. Shoreline response to a sequence of typhoon and monsoon events. Water. 2017;**9**:364

[8] Wu Z, Jiang C, Deng B, Chen J, Li L. Evaluation of numerical wave model for typhoon wave simulation in South China Sea. Water Science and Engineering. 2018;**11**:229-235

[9] Emanuel K. Increasing destructiveness of tropical cyclones over the past 30 years. Nature. 2005;**436**:686

[10] Webster PJ, Holland GJ, Curry JA. Changes in tropical cyclone number, duration, and intensity in a warming environment. Science. 2005;**309**:1844-1846

[11] Yin X, Wang Z, Liu Y. Ocean response to typhoon Ketsana traveling over the Northwest Pacific and a numerical model approach. Geophysical Research Letters. 2007;**34**:21606

[12] Chen W-B, Lin L-Y, Jang J-H, Chang C-H. Simulation of typhoon-induced storm tides and wind waves for the northeastern coast of Taiwan using a tide–surge–wave coupled model. Water. 2017;**9**:549

[13] Potter H, Drennan WM, Graber HC. Upper Ocean cooling and air-sea fluxes under typhoons: A case study. Journal of Geophysical Research, Oceans. 2017;**122**:7237-7252

[14] Li ZL, Wen P. Comparison between the response of the Northwest Pacific Ocean and the South China Sea to typhoon Megi (2010). Advances in Atmospheric Sciences. 2017;**34**:79-87

[15] Chen Y, Yu X. Enhancement of wind stress evaluation method under storm conditions. Climate Dynamics. 2016;**47**:3833-3843

[16] Chen Y, Yu X. Sensitivity of storm wave modeling to wind stress evaluation methods. Journal of Advances in Modeling Earth Systems. 2017;**9**:893-907

[17] Fan Y, Ginis I, Hara T. The effect of wind–wave–current interaction on air–sea momentum fluxes and ocean response in tropical cyclones. Journal of Physical Oceanography. 2009;**39**:1019-1034

[18] Gronholz A, Gräwe U, Paul A. Investigating the effects of a summer storm on the North Sea stratification using a regional coupled

ocean-atmosphere model. Ocean Dynamics. 2017;**67**:1-25

[19] Mattocks C, Forbes C. A real-time, event-triggered storm surge forecasting system for the state of North Carolina. Ocean Modelling. 2008;**25**:95-119

[20] Xu S, Huang W, Zhang G. Integrating Monte Carlo and hydrodynamic models for estimating extreme water levels by storm surge in Colombo, Sri Lanka. Natural Hazards. 2014;**71**:703-721

[21] Zhang K, Li Y, Liu H. Transition of the coastal and estuarine storm tide model to an operational storm surge forecast model: A case study of the Florida coast. Weather and Forecasting. 2013;**28**:1019-1037

[22] Black PG, D'Asaro EA, Sanford TB. Air-sea exchange in hurricanes: Synthesis of observations from the coupled boundary layer air–sea transfer experiment. Bulletin of the American Meteorological Society. 2007;**88**:357-374

[23] Chen Y, Zhang F, Green BW. Impacts of ocean cooling and reduced wind drag on hurricane Katrina (2005) based on numerical simulations. Monthly Weather Review. 2018;**146**:287-306

[24] Liu B, Liu H, Xie L. A coupled atmosphere-wave-ocean modeling system: Simulation of the intensity of an idealized tropical cyclone. Monthly Weather Review. 2010;**139**:132-152

[25] Mori N, Kato M, Kim S. Local amplification of storm surge by super typhoon Haiyan in Leyte gulf. Geophysical Research Letters. 2014;**41**:5106-5113

[26] Takagi H, Esteban M, Shibayama T. Track analysis, simulation, and field survey of the 2013 typhoon Haiyan storm surge. Journal of Flood Risk Management. 2017;**10**:42-52

[27] Zhang Y, Liu L, Bi S, Wu Z, Shen P, Ao Z, et al. Analysis of dual-polarimetric radar variables and quantitative precipitation estimators for landfall typhoons and squall lines based on disdrometer data in southern China. Atmosphere. 2019;**10**:30

[28] Huang C, Hu J, Chen S, Zhang A, Liang Z, Tong X, et al. How well can IMERG products capture typhoon extreme precipitation events over southern China? Remote Sensing. 2019;**11**:70

[29] Guo Y, Zhang J, Zhang L, Shen Y. Computational investigation of typhoon-induced storm surge in Hangzhou Bay, China. Estuarine, Coastal and Shelf Science. 2009;**85**:530-536

[30] Yin K, Xu S, Huang W. Effects of sea level rise and typhoon intensity on storm surge and waves in Pearl River estuary. Ocean Engineering. 2017;**136**:80-93

[31] Wang J, Yi S, Li M. Effects of sea level rise, land subsidence, bathymetric change and typhoon tracks on storm flooding in the coastal areas of Shanghai. Science of the Total Environment. 2018;**621**:228-234

[32] Wu Z, Jiang C, Deng B, Chen J, Long Y, Qu K, et al. Numerical investigation of typhoon Kai-tak (1213) using a mesoscale coupled WRF-ROMS model. Ocean Engineering. 2019;**175**:1-15

[33] Shen Y, Jia H, Li C, Tang J. Numerical simulation of saltwater intrusion and storm surge effects of reclamation in Pearl River estuary, China. Applied Ocean research. 2018;**79**:101-112

[34] Charnock H. Wind stress on a water surface. Quarterly Journal of the Royal Meteorological Society. 1955;**81**:639-640

[35] Taylor PK, Yelland MJ. The dependence of sea surface roughness on

the height and steepness of the waves. Journal of Physical Oceanography. 2001;**31**:572-590

[36] Wu Z, Jiang C, Chen J, Long Y, Deng B, Liu X. Three-dimensional temperature field change in the South China Sea during typhoon Kai-Tak (1213) based on a fully coupled atmosphere–wave–ocean model. Water. 2019;**11**:140

[37] Bilskie M, Hagen S, Medeiros S, Cox A, Salisbury M, Coggin D. Data and numerical analysis of astronomic tides, wind-waves, and hurricane storm surge along the northern Gulf of Mexico. Journal of Geophysical Research, Oceans. 2016;**121**:3625-3658

[38] Fakour H, Lo S, Lin T. Impacts of typhoon Soudelor (2015) on the water quality of Taipei, Taiwan. Scientific Reports. 2016;**6**:25228

[39] Wu Z, Jiang C, Deng B, Cao Y. Simulation of the storm surge in the South China Sea based on the coupled sea-air model. Chinese Science Bulletin. 2018;**63**:3494-3504

[40] Wang T, Liu G, Gao L, Zhu L, Fu Q, Li D. Biological and nutrient responses to a typhoon in the Yangtze estuary and the Adjacent Sea. Journal of Coastal Research. 2015;**32**:323-332

[41] Chen Z, Pan J, Jiang Y, Lin H. Far-reaching transport of Pearl River plume water by upwelling jet in the northeastern South China Sea. Journal of Marine Systems. 2017;**173**:60-69

[42] Zhang Z, Wu H, Yin X, Qiao F. Dynamics response of Changjiang River plume to a severe typhoon with the surface wave-induced mixing. Journal of Geophysical Research, Oceans. 2018;**123**:9369-9388. DOI: 10.1029/2018JC014266

[43] Hu J, Wang X. Progress on upwelling studies in the China seas. Reviews of Geophysics. 2016;**54**:653-673

[44] Tsai Y, Chung C, Chung C, Gau H, Lai W, Liao S. The impact of typhoon Morakot on heavy metals of Dapeng Bay and pollution from neighboring Rivers. Environmental Modeling and Assessment. 2016;**21**:479-487

[45] Jiang C, Wu Z, Chen J, Deng B, Long Y. Sorting and sedimentology character of sandy beach under wave action. Procedia Engineering. 2015;**116**:771-777

[46] Jiang C, Wu Z, Chen J, Deng B, Long Y, Li L. An available formula of the sandy beach state induced by plunging waves. Acta Oceanologica Sinica. 2017;**36**:91-100

[47] Huang W. Modelling the effects of typhoons on morphological changes in the estuary of Beinan, Taiwan. Continental Shelf Research. 2017;**135**:1-13

[48] Wu X, Wang H, Bi N, Song Z, Zang Z, Kineke G. Bio-physical changes in the coastal ocean triggered by typhoon: A case of typhoon Meari in summer 2011. Estuarine, Coastal and Shelf Science. 2016;**183**:413-421

[49] Wang T, Liu G, Gao L, Zhu L, Fu Q, Li D. Biological responses to nine powerful typhoons in the East China Sea. Regional Environmental Change. 2017;**17**:465-476

[50] Dai Z, Mei X, Darby S, Lou Y, Li W. Fluvial sediment transfer in the Changjiang (Yangtze) river-estuary depositional system. Journal of Hydrology. 2018;**566**:719-734

[51] Chen N, Krom M, Wu Y, Yu D, Hong H. Storm induced estuarine turbidity maxima and controls on nutrient fluxes across river-estuary-coast continuum. Science of the Total Environment. 2018;**628**:1108-1120

[52] Xu J, Wang N, Li G, Dong P, Li J, Liu S, et al. The dynamic responses

of flow and near-bed turbidity to typhoons on the continental shelf of the East China Sea: Field observations. Geological Journal. 2016;**51**:12-21

[53] Bernardo L, Nadaoka K, Nakamura T, Watanabe A. Island-enhanced cooling mechanism in typhoon events revealed by field observations and numerical simulations for a coral reef area, Sekisei lagoon, Japan. Ocean Dynamics. 2017;**67**:1369-1384

[54] Zhang H, Chen D, Zhou L, Liu X, Ding T, Zhou B. Upper Ocean response to typhoon Kalmaegi (2014). Journal of Geophysical Research, Oceans. 2016;**121**:6520-6535

[55] Sun J, Oey L, Chang R, Xu F, Huang S. Ocean response to typhoon Nuri (2008) in western Pacific and South China Sea. Ocean Dynamics. 2015;**65**:735-749

[56] Lin I. Typhoon-induced phytoplankton blooms and primary productivity increase in the western North Pacific subtropical ocean. Journal of Geophysical Research, Oceans. 2012;**117**:C03039

[57] Liu F, Tang S. Influence of the interaction between typhoons and oceanic Mesoscale eddies on phytoplankton blooms. Journal of Geophysical Research, Oceans. 2018;**123**:2785-2794

[58] Li T, Bai Y, He X, Chen X, Chen C-TA, Tao B, et al. The relationship between POC export efficiency and primary production: Opposite on the shelf and basin of the northern South China Sea. Sustainability. 2018;**10**:3634

[59] Tsai Y, Chern C, Jan S, Wang J. Numerical study of cold dome variability induced by typhoon Morakot (2009) off northeastern Taiwan. Journal of Marine Research. 2013;**71**:109-131

[60] Li D, Huang C. The influences of orography and ocean on track of typhoon Megi (2016) past Taiwan as identified by HWRF. Journal of Geophysical Research-Atmospheres. 2018;**123**:11,492-11,517

[61] Wu R, Li C. Upper ocean response to the passage of two sequential typhoons. Deep-Sea Research Part I. 2018;**132**:68-79

[62] Yue X, Zhang B, Liu G, Li X, Zhang H, He Y. Upper ocean response to typhoon Kalmaegi and Sarika in the South China Sea from multiple-satellite observations and numerical simulations. Remote Sensing. 2018;**10**:348

[63] Pujiana K, Moum J, Smyth W. The role of turbulence in redistributing upper-ocean heat, freshwater, and momentum in response to the MJO in the equatorial Indian ocean. Journal of Physical Oceanography. 2018;**48**:197-220

[64] Zhang X, Chu P, Li W, Liu C, Zhang L, Shao C, et al. Impact of langmuir turbulence on the thermal response of the ocean surface mixed layer to Supertyphoon Haitang (2005). Journal of Physical Oceanography. 2018;**48**:1651-1674

[65] Warner JC, Sherwood CR, Signell RP. Development of a three-dimensional, regional, coupled wave, current, and sediment-transport model. Computational Geosciences. 2008;**34**:1284-1306

[66] Warner JC, Armstrong B, He R. Development of a coupled ocean–atmosphere–wave–sediment transport (COAWST) modeling system. Ocean Modelling. 2010;**35**:230-244

[67] Balaguru K, Foltz G, Leung L, Emanuel K. Global warming-induced upper-ocean freshening and the intensification of super typhoons. Nature Communications. 2016;7:13670

[68] Lee J, Moon I, Moon J, Kim S, Jeong Y, Koo J. Impact of typhoons on the Changjiang plume extension in the yellow and East China seas. Journal of Geophysical Research, Oceans. 2017;**122**:4962-4973

[69] Nakada S, Kobayashi S, Hayashi M, Ishizaka J, Akiyama S, Fuchi M, et al. High-resolution surface salinity maps in coastal oceans based on geostationary ocean color images: Quantitative analysis of river plume dynamics. Journal of Oceanography. 2018;**74**:287-304

[70] Rudzin J, Shay L, Johns W. The influence of the barrier layer on SST response during tropical cyclone wind forcing using idealized experiments. Journal of Physical Oceanography. 2018;**48**:1951-1968

[71] Lengaigne M, Neetu S, Samson G, Vialard J, Krishnamohan K, Masson S, et al. Influence of air–sea coupling on Indian ocean tropical cyclones. Climate Dynamics. 2019;**52**:577-598

[72] Steffen J, Bourassa M. Barrier layer development local to tropical cyclones based on Argo float observations. Journal of Physical Oceanography. 2018;**48**:1951-1968

[73] Wu Z, Chen J, Jiang C, et al. Numerical investigation of Typhoon Kai-tak (1213) using a mesoscale coupled WRF-ROMS model—Part II: Wave effects[J]. Ocean Engineering; 2020;**196**:106805

[74] Blumberg AF, Georgas N, Yin L, et al. Street-scale modeling of storm surge inundation along the New Jersey Hudson river waterfront. Journal of Atmospheric and Oceanic Technology. 2015;**32**(8):1486-1497

[75] Blumberg AF, Mellor GL. A description of a three-dimensional coastal ocean circulation model. In: Three-Dimensional Coastal Ocean Models. Washington, USA; 1987. pp. 1-16

[76] Chen T, Zhang Q, Wu Y, et al. Development of a wave-current model through coupling of FVCOM and SWAN. Ocean Engineering. 2018;**164**:443-454

[77] Chen S, Qian YK, Peng S. Effects of various combinations of boundary layer schemes and microphysics schemes on the track forecasts of tropical cyclones over the South China Sea. Natural Hazards. 2015;**78**(1):61-74

[78] Craig AP, Jacob R, Kauffman B, et al. CPL6: The new extensible, high performance parallel coupler for the community climate system model. The International Journal of High Performance Computing Applications. 2005;**19**(3):309-327

[79] Di Liberto T, Colle BA, Georgas N, et al. Verification of a multimodel storm surge ensemble around new York City and Long Island for the cool season. Weather and Forecasting. 2011;**26**(6):922-939

[80] Forbes C, Rhome J, Mattocks C, et al. Predicting the storm surge threat of hurricane Sandy with the National weather service SLOSH model. Journal of Marine Science and Engineering. 2014;**2**(2):437-476

[81] Ge Z, Dai Z, Pang W, et al. LIDAR-based detection of the post-typhoon recovery of a meso-macro-tidal beach in the Beibu gulf, China. Marine Geology. 2017;**391**:127-143

[82] Haidvogel DB, Arango H, Budgell WP, et al. Ocean forecasting in terrain-following coordinates: Formulation and skill assessment of the regional ocean modeling system. Journal of Computational Physics. 2008;**227**(7):3595-3624

[83] Hu D, Wu L, Cai W, et al. Pacific western boundary currents and their roles in climate. Nature. 2015;**522**(7556):299

[84] Hu K, Chen Q, Wang H. A numerical study of vegetation impact on reducing storm surge by wetlands in a semi-enclosed estuary. Coastal Engineering. 2015;**95**:66-76

[85] Islam T, Srivastava PK, Rico-Ramirez MA, et al. Tracking a tropical cyclone through WRF–ARW simulation and sensitivity of model physics. Natural Hazards. 2015;**76**(3):1473-1495

[86] Jelesnianski CP, Chen J, Shaffer WA. SLOSH: Sea, lake, and overland surges from hurricanes. In: National Weather Service. USA; 1992

[87] Jones PW. A Users Guide for SCRIP: A Spherical Coordinate Remapping and Interpolation Package. Los Alamos National Laboratory; 1998. Available from: http://climate.lanl.gov/Software/SCRIP/

[88] Lakshmi DD, Murty PLN, Bhaskaran PK, et al. Performance of WRF-ARW winds on computed storm surge using hydodynamic model for Phailin and Hudhud cyclones. Ocean Engineering. 2017;**131**:135-148

[89] Laprise R. The Euler equations of motion with hydrostatic pressure as an independent variable. Monthly Weather Review. 1992;**120**(1):197-207

[90] Li M, Zhong L, Boicourt WC, et al. Hurricane-induced storm surges, currents and destratification in a semi-enclosed bay. Geophysical Research Letters. 2006;**33**(2):L02604

[91] Li Y, Peng S, Yan J, et al. On improving storm surge forecasting using an adjoint optimal technique. Ocean Modelling. 2013;**72**:185-197

[92] Lorbacher K, Dommenget D, Niiler PP. Ocean mixed layer depth: A subsurface proxy of ocean-atmosphere variability. Journal of Geophysical Research, Oceans. 2006;**111**:520-522

[93] Liu D, Pang L, Xie B. Typhoon disaster in China: Prediction, prevention, and mitigation. Natural Hazards. 2009;**49**(3):421-436

[94] Liu N, Ling T, Wang H, et al. Numerical simulation of typhoon Muifa (2011) using a coupled ocean-atmosphere-wave-sediment transport (COAWST) modeling system. Journal of Ocean University of China. 2015;**14**(2):199-209

[95] Mei X, Dai Z, Darby SE, et al. Modulation of extreme flood levels by impoundment significantly offset by floodplain loss downstream of the three gorges dam. Geophysical Research Letters. 2018;**45**(7):3147-3155

[96] Mooney PA, Mulligan FJ, Bruyère CL, et al. Investigating the performance of coupled WRF-ROMS simulations of hurricane Irene (2011) in a regional climate modeling framework. Atmospheric Research. 2019;**215**:57-74

[97] Neumann JE, Emanuel K, Ravela S, et al. Joint effects of storm surge and sea-level rise on US coasts: New economic estimates of impacts, adaptation, and benefits of mitigation policy. Climatic Change. 2015;**129**(1-2):337-349

[98] Pattanayak S, Mohanty UC, Rao AD. Simulation of storm surges in the bay of Bengal using one-way coupling between NMM-WRF and IITD storm surge model. Marine Geodesy. 2016;**39**(5):376-400

[99] Rego JL, Li C. Nonlinear terms in storm surge predictions: Effect of tide and shelf geometry with case study from hurricane Rita. Journal of Geophysical Research, Oceans. 2010;**115**(C6). DOI: 10.1029/2009JC005285

Chapter 5

The Use of a Spiral Band Model to Estimate Tropical Cyclone Intensity

Boris Yurchak

Abstract

Spiral cloud-rain bands (SCRBs) are some of the most distinguishing features inherent in satellite and radar images of tropical cyclones (TC). The subject of the proposed research is the finding of a physically substantiated method for estimation of the TC's intensity using SCRBs' configuration parameters. To connect a rainband pattern to a physical process that conditions the spiraling feature of a rainband, it is assumed that the rainband's configuration near the core of a TC is governed primarily by a streamline. In turn, based on the distribution of primarily forces in a TC, an analytical expression as a combination of hyperbolic and logarithmic spirals (HLS) for the description of TC spiral streamline (rainband) is retrieved. Parameters of the HLS are determined by the physical parameters of a TC, particularly, by the maximal wind speed (MWS). To apply this theoretical finding to practical estimation of the TC's intensity, several approximation techniques are developed to "convert" rainband configuration to the estimation of the MWS. The developed techniques have been tested by exploring satellite infrared imageries and airborne and coastal radar data, and the outcomes were compared with in situ measurements of wind speeds and the best track data of tropical cyclones.

Keywords: hyperbolic-logarithmic spiral, tropical cyclone, spiral cloud-rain bands, maximum wind speed, approximation

1. Introduction

The issue addressed in this chapter relates to methods for estimating the intensity of a tropical cyclone (TC) from the characteristics of its cloud-rain field (CRF) structure. In general, these methods are empirical and semiempirical, i.e., they are based on the correlation of the structural features of the CRF with the intensity of the TC found from observations from satellites and radars. The most widely used method in operational practice is the Dvorak method [1, 2]. This chapter relates to the exploring of one of the most pronounced structural elements of the CRF, which are spiral cloud-rain bands (SCRBs). Attention was first attracted to these bands by Wexler [3] based on aircraft observations. It was suggested that SCRBs indicate the mature cyclone and its organization follows the streamlines. The authors of [4] described the SCRBs observed on the radar and suggested to use a modified logarithmic spiral to express its configuration in mathematic form. Although SCRBs have been studied for a long time, there is currently no consensus about their origin

 IntechOpen

and mechanism of generation. Reviews of proposed hypotheses are given in [5–8]. In the paper of Lahiri [9], a first attempt was undertaken to estimate the effect of the TC intensity on the geometric characteristics of SCRBs. In this study, a simple model of a TC in which the low-level streamlines were described by logarithmic spirals was suggested. It was shown that the rate of generation of a latent heat in the model was proportional to the crossing angle that is a parameter of the spiral. (As follows from [4], the "crossing angle" of a SCRB of a finite width is the angle at which the longitudinal axial curve of the SCRB crosses the concentric circle (centered at the center of the cyclone) at a given arbitrary point belonging to the axial curve; that is, this is the angle between the tangents to the axial curve and the said circle at this point). As a result, crossing angle decreases as cyclones matured since the latent heat is very small in this stage. However, no relationship between the maximum wind speed (MWS) and the crossing angle was obtained. Moreover, the description of SCRBs by a logarithmic spiral only is an internally contradictory approximation due to the following circumstances. On the one hand, the main property of the logarithmic spiral is the constancy of the crossing angle for any of its points. Therefore, in particular, the alternative name of this spiral is an equiangular spiral. On the other hand, it was found that the crossing angle is sensitive to wind speed. But wind speed is not constant along the SCRB. Therefore, the crossing angle cannot have a constant value along the SCRB as well. Thus, the main feature of the logarithmic spiral (constancy of the crossing angle) is not consistent with the physics of the process. In the modified logarithmic spiral [4], the experimentally observed dependence of the crossing angle on distance is taken into account by introducing a radius of so-called inner limiting circle, at which the crossing angle is zero. However, the relationship of this radius with the intensity of the cyclone and the radius of the maximum wind (RMW) has not been established. It should be noted that the "spiralness" of the SCRB is also used in the empirical Dvorak method, although only at the qualitative level, by estimating the sector occupied by the spiral structure ("count the tenths" method), which is approximated by a logarithmic spiral with the crossing angle of 10°. However, as it was stated in [10], this spiral does not have a physical basis. In our papers [11–14], the assumption of authors [3, 4, 9] that rainbands are well arranged along streamlines was also used. In general, this assumption is confirmed by a comparison of radar and aircraft data (e.g., [15, 16]). The same orientation of the principal rainband along the jet is demonstrated in [7]. Further, the expression for spiral streamline has been derived in [11] in the closed form as hyperbolic-logarithmic spiral (HLS). The most advantage feature of the HLS is the dependence of one of its parameters on the MWS. Unlike the previous studies, the HLS was not assumed but was accurately derived based on physical considerations. At the same time, it turned out that only the peripheral portion of the HLS is similar to the logarithmic spiral. The size of this portion and the corresponding crossing angle are determined by the parameters of the cyclone, including the MWS. Unlike the modified logarithmic spiral, the change in the tangent of the crossing angle in the HLS is governed not by the distance weight function (equal to zero at the distance equal to the radius of the inner limiting circle and approaches unity at the outer edge of the cyclone), but by the physical parameters of the cyclone. However, it is not possible to determine the MWS by the crossing angle of the logarithmic section of the HLS only, as well, due to multifactor influence on the crossing angle. On the other hand, approximation of a SCRB by the HLS allows determining the MWS based on the known characteristics of the cyclone in cases when the SCRB is sufficiently long and markedly different from the logarithmic spiral. A discussion of the physical basis of the proposed method, methodology, and results of its application is the subject of this chapter. In particular, the derivation of the HLS based on the distribution of forces affecting the cyclone, methods for approximating the SCRB, the results of estimating the MWS from radar, and satellite data in comparison with the data of direct measurements and final conclusions of the corresponding meteorological services (Best Track reports) are provided.

2. Hyperbolic-logarithmic model of a streamline in a cyclone

2.1 Streamline equation in polar coordinates

As per Batchelor [17], radial (v) and tangential (u) components of a fluid flow are respectively:

$$v = \frac{1}{r}\frac{d\psi}{d\varphi}, \tag{1}$$

$$u = -\frac{d\psi}{dr}, \tag{2}$$

where r and φ are polar radius and polar angle, respectively, and Ψ is the stream function. The combination of (1) and (2) results the streamline equation in polar coordinates:

$$-r\frac{d\varphi}{u} = \frac{1}{v}dr. \tag{3}$$

Considering $\frac{u}{v} = \tan(\mu)$, where μ is the inflow angle (**Figure 1**), the final form of the streamline equation is

$$-\frac{d\varphi}{dr} = \frac{1}{r}\tan\mu(r) \tag{4}$$

It should be noted that the sum of inflow and crossing angles is the right angle (**Figure 1**).

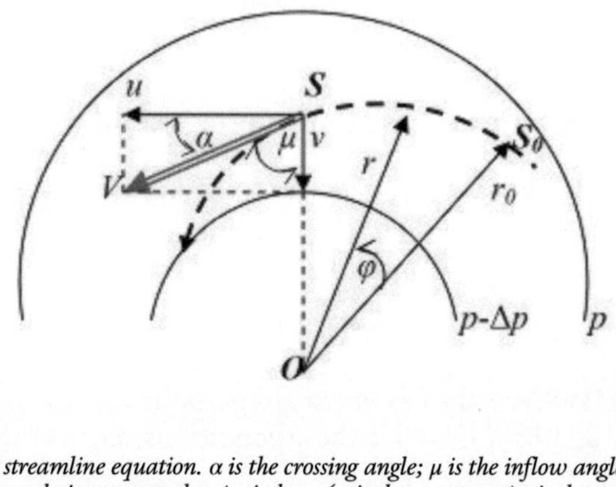

Figure 1.
To the derivation of the streamline equation. α is the crossing angle; μ is the inflow angle; the bold dashed curve is the streamline; solid curves designate p and p-Δp isobars (p is the pressure, Δp is the pressure change between isobars); V is the wind speed of air in point S; u and v are tangential and radial components of the wind speed V at point S, respectively; and φ is the polar angle of radius vector r counted off from arbitrary selected point S_o.

2.2 Inflow angle from the balance of forces in a cyclone

To express the streamline Eq. (4) through the physical parameters of a cyclone, the inflow angle should be evaluated based on the balance of forces in a cyclone. The diagram of forces is depicted in **Figure 2** as follows from [18].

The balance condition called Guldberg-Mohn balance [19] is expressed as follows:

$$F_G = F_B \tag{5}$$

where

$$F_G = -\frac{1}{\rho}\frac{\partial p}{\partial r} \tag{6}$$

is the gradient force (ρ is the air density);

$$\vec{F_B} = \vec{F_c} + \vec{F_d} + \vec{F_R} \tag{7}$$

is the balancing force containing

$$F_c = \frac{V^2}{r}, \tag{8}$$

which is the centrifugal force;

$$F_d = V \cdot f, \tag{9}$$

which is the deflecting force ($f = 2\omega \sin\phi$ is the Coriolis parameter, ω is the angular speed of the Earth rotation, and ϕ is the altitude); and

$$F_R = -k \cdot V, \tag{10}$$

which is the frictional force, where k is the friction factor (s^{-1}). All forces in the above expressions (8)–(10) are given per unit mass. As it follows from **Figure 2** and Eqs. (8)–(10), the tangent of inflow angle under the steady-state air movement in the friction layer is defined by a relationship:

$$\tan\mu(r) = \frac{F_c + F_d}{|F_R|} = \frac{f}{k} + \frac{V(r)}{kr}. \tag{11}$$

2.3 The streamline equation as a function of physical parameters of a cyclone

Let us assume for definiteness that the speed of wind $V(r)$ changes with cyclone radius in accordance with a power law which is inherent to the Rankine vortex [20]:

$$V(r) = V_{\mathrm{m}}\left(\frac{r_{\mathrm{m}}}{r}\right)^n, \quad r \geq r_{\mathrm{m}}, \tag{12}$$

where V_{m} is the MWS, n is the exponent (hyperbolic index), and r_{m} is the radius of the maximum wind (RMW). Under these conditions, Eq. (4) can be written as

$$-d\varphi = \left(\frac{B}{r} + b\frac{1}{r^{n+2}}\right)dr, \tag{13}$$

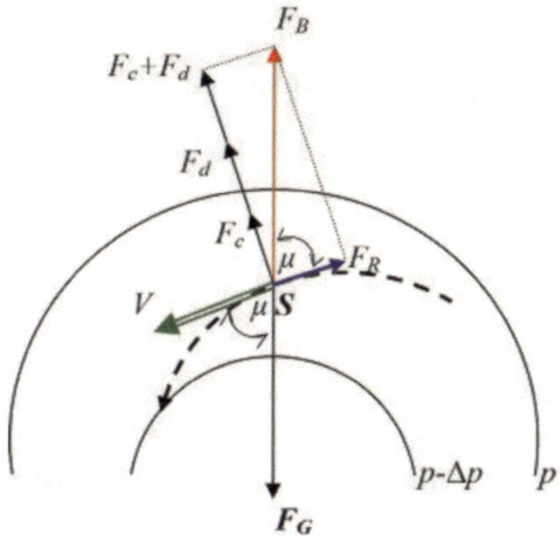

Figure 2.
Balance of forces in a cyclone under steady-state air movement in the friction layer (after [18]); designations are the same as for ***Figure 1.***

where

$$B = f/k \tag{14}$$

and

$$b = V_m r_m^n / k. \tag{15}$$

The solution of Eq. (13) is [11]

$$-\varphi = B\ln\frac{r}{r_0} - \frac{b}{n+1}\left(\frac{1}{r^{n+1}} - \frac{1}{r_0^{n+1}}\right), \tag{16}$$

where r_0 is the radius where it is assumed that $\varphi = 0$. Simplifying this expression, one can get

$$\varphi = A \cdot \left(\frac{1}{y^{n+1}} - 1\right) - B \cdot \ln y, \tag{17}$$

where $y = r/r_0$ is the normalized polar radius and coefficient A is given by an expression

$$A = \frac{r_m^n}{k(n+1)r_0^{n+1}} V_m. \tag{18}$$

The first and second terms of Eq. (17) can be named, respectively, as hyperbolic and logarithmic components of the streamline equation of wind in a TC in the polar coordinates. Accordingly, Eq. (17) can be named the hyperbolic-logarithmic spiral.

2.4 Alternate representations of HLS streamline

Because $y^{n+1} \equiv \exp\{(n+1)\ln y\}$, the expression (17) can be transformed to the exponent-logarithmic form

$$\phi = A \cdot \left\{e^{-(n+1)\ln y} - 1\right\} - B \cdot \ln y. \tag{19}$$

Using the Maclaurin series expansion for an exponential function $\left(e^x = \sum_{i=0}^{\infty} x^i/i!\right)$ and combining the coefficients at the argument of the first power, Eq. (19) can be written as a polynomial:

$$\varphi = A\sum_{i=2}^{\infty}\frac{\{-(n+1)\}^i}{i!}(\ln y)^i - \{A(n+1)+B\}\ln y \tag{20}$$

As follows from Eq. (20), the HLS contains linear and nonlinear parts in regard to the logarithm of the relative polar radius ($\ln y$). Taking into account the expression for coefficient A (18), the linear part is the logarithmic spiral (φ_L) that can be expressed in the form:

$$\varphi_L = -\{A(n+1)+B\}\ln y = -\left(\frac{r_m^n}{kr_0^{n+1}}V_m + B\right)\ln y = -B\left(y_m^n\frac{V_m}{V_C}+1\right)\ln y \tag{21}$$

where $V_C = fr_0$ is the Coriolis speed at distance r_0 and $y_m = r_m/r_0$ is the relative RMW. As follows from Eq. (21), the tangent of the crossing angle of the linear part of the HLS depends on many factors as follows:

$$\tan\alpha = \left\{B\left(y_m^n\frac{V_{max}}{V_C}+1\right)\right\}^{-1}. \tag{22}$$

2.5 Graphical representations of HLS streamline

To illustrate the dependence of HLS primary features versus the MWS and the friction factor, the graphical diagrams in polar coordinates (**Figure 3**) and semilogarithmic coordinates (**Figure 4**) for three values of MWS and two values of the friction factor are provided below.

Figure 3 reflects the known feature of a rainband, as it has been summarized by Willoughby [16]. A rainband of a cyclone from the periphery to its center is curved, due to the increase in the angular velocity, and therefore can be transformed into a logarithmic spiral. As the intensity of a cyclone increases, the slope of the twist decreases, and the band configuration becomes like a circular arc. Indeed, the HLS is similar to a logarithmic spiral, but only on the periphery of a cyclone, as shown in Eq. (20), where the normalized polar radius (y) is a little less than 1. Toward the center of the cyclone ($y < 1$), the first term in Eq. (20) begins to prevail over the second one, and the HLS begins to "round off" (with constant k and n) accordingly with the observations. The distance where it happens depends on the parameter ratio between the linear and nonlinear parts of Eq. (20). Plots of HLS with V_m = 40 m s^{-1} are missing in **Figure 4** to simplify its reading. In the semilogarithmic coordinates in **Figure 4**, the canonical logarithmic spiral $\varphi = -(\tan\alpha)^{-1}\cdot\ln y$ would be depicted as a straight line with the angle coefficient equal to the inverse value of the tangent of the corresponding crossing angle. As follows from the plots in **Figure 4**, the HLS is more similar to the logarithmic spiral (straight line) within one turn (2π), and for $0.1 \le y \le 1$ ($-2.3 \le \ln y \le 0$) for any MWS, the greater is the friction factor (compare lower and upper pairs of curves). For small friction factor, the greater the MWS is, the shorter the linear part of the HLS (logarithmic) is (compare two lower curves). As follows from Eq. (22), in contrast to the canonical logarithmic spiral mentioned above where $\tan\alpha = const$ at any y, the tangent of the crossing angle of the HLS linear part depends on the term $y_m^n V_m/V_C$. It is a reason

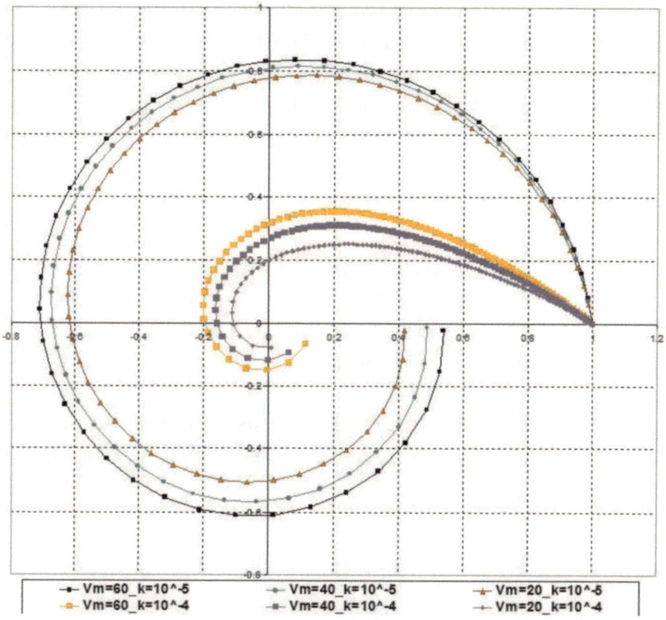

Figure 3.
Illustration of HLS configuration in polar coordinates versus variable maximal wind speed and friction factor; ordinate and abscissa axis are relative polar radius (y); $V_m(m\,s^{-1})$, $k\,(s^{-1})$, $r_o = 500$ *km,* $n = 0.6$, $y_m = 0.1$, *and* $f = 1{\cdot}10^{-5}\,s^{-1}$ *(latitude = 20°N).*

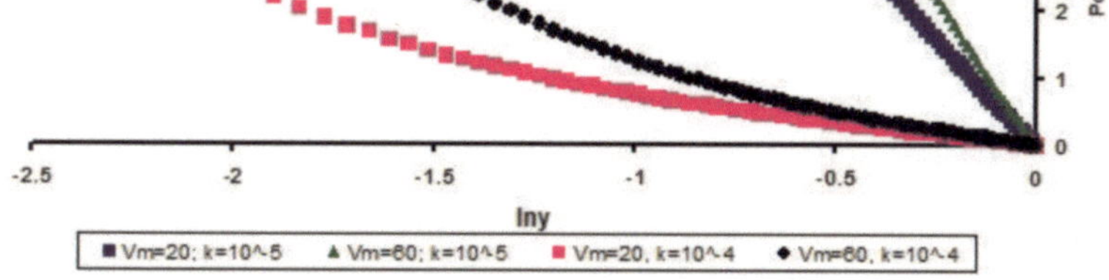

Figure 4.
Illustration of HLS configuration in semilogarithmic coordinates versus variable maximal wind speed and friction factor; $V_m(m\,s^{-1})$, $k\,(s^{-1})$, $r_o = 500$ *km,* $n = 0.6$, $y_m = 0.1$, *and* $f = 1{\cdot}10^{-5}\,s^{-1}$ *(latitude = 20° N).*

why the linear parts of pairs of curves in **Figure 4** with the same friction factor ($k = 10^{-5}\,s^{-1}$ or $k = 10^{-4}\,s^{-1}$) and the same Coriolis parameter (f) have notable different tilts for different MWS. It is the explanation of the well-known experimental fact, mentioned in the Introduction section, regarding the sensitivity of the crossing angle on the intensity of a TC.

3. HLS approximation techniques

The task of using the analytical expression of the streamline (17) or (19) considered in the previous section is to use it to determine the MWS in a TC. The base for resolving this task is the expression (18) for the coefficient A that included the MWS. From this expression, taking into account the relationship between the friction coefficient and the coefficient B (14), one can get the desired formulae for calculation the MWS using the HLS parameters:

$$V_{\mathrm{m}} = A\frac{k(n+1)r_0^{n+1}}{r_{\mathrm{m}}^n} = \frac{A}{B}\frac{f(n+1)r_0^{n+1}}{r_{\mathrm{m}}^n} = \frac{A}{B}(n+1)y_{\mathrm{m}}^{-1}V_{\mathrm{C}} \tag{23}$$

Thus, the MWS can be estimated by approximating the SCRB of the TC with the HLS Eq. (19). Under this approximation, values of A and B are determined, assuming the value of the hyperbolic index n, as well as the measured or estimated value of the MWR. Coriolis velocity (V_{C}) is determined by known latitude of the center of the TC and the radius of given initial point of the SCRB. To get the approximation estimate of the HLS coefficients, the standard least squares method (LSM) and the assimilation technique were considered and tested. Brief descriptions of these techniques are provided below.

3.1 Estimation of coefficients of the HLS by the least squares method

3.1.1 Relationships for calculation

Taking the exponentially logarithmic form of the HLS (19) and denoting $\ln y = x$ and $T(x) = e^{-(n+1)x} - 1$, it is possible to write

$$\varphi = A \cdot T(x) - B \cdot x. \tag{24}$$

Performing the routine LSM procedure, one gets the calculation relationships for the HLS coefficients estimates which are provided below in the Gauss designation ($[s] = \sum s$)

$$\tilde{A} = \frac{1}{D}\left\{ [x^2] \cdot [\varphi \cdot T(x)] - [x \cdot T(x)] \cdot [\varphi \cdot x] \right\}, \tag{25}$$

$$\tilde{B} = \frac{1}{D}\left\{ [x \cdot T(x)] \cdot [\varphi \cdot T(x)] - [T^2(x)] \cdot [\varphi \cdot x] \right\}. \tag{26}$$

where

$$D = [T^2(x)][x^2] - [x \cdot T(x)]^2. \tag{27}$$

The error estimates of the coefficients are equal:

$$\sigma_A = \sigma \cdot \sqrt{Q_{11}}, \tag{28}$$

$$\sigma_B = \sigma \cdot \sqrt{Q_{22}}, \tag{29}$$

where

$$\sigma = \sqrt{\frac{[\Delta\varphi^2]}{N-2}}, \tag{30}$$

is the residual variance, where

$$\left[\Delta\varphi^2\right] = \sum_{i=1}^{N}\left(\varphi_{i,\,\exp er.} - \varphi_{i,\,approx.}\right)^2 = \sum_{i=1}^{N}\left(\varphi_{i,\,\exp er.} - \tilde{A}\cdot T(x_i) + \tilde{B}\cdot x_i\right)^2 \quad (31)$$

and terms in (28) and (29) are $Q_{11} = \frac{[x^2]}{D}$, $Q_{22} = \frac{[T^2(x)]}{D}$.

3.1.2 The condition of stationarity under the application of the LSM for approximation of a SCRB with the HLS

As follows from Eq. (18), coefficient A of the HLS depends on the selection of radius r_0. On the other hand, based on the physical sense, the cyclone's physical parameters evaluated from coefficients A and B must not depend on selection of the initial point (point of spiral reference). To satisfy this requirement, the selected spiral signature should be fitted by the HLS with coefficient A that should be proportional to the polar radius with power $-(n + 1)$ (in accordance with Eq. (18)) and coefficient B that should be a constant (in accordance with (14)). This validates the fitting of a given spiral signature by the HLS. Substantiation of these conditions, which characterize the "HLS stationarity" and the control technique, is provided below. The term "HLS stationarity" can be explained by providing the following example. Suppose some spiral signature is introduced by a set of point coordinates. This set of points is aimed to be approximated by the HLS using the LSM. Let us select arbitrary the first starting point of approximation (SPA) at the peripheral part of the spiral signature. Suppose its polar radius is $r_{0,1}$. The applied LSM procedure evaluates parameters A_1 and B_1. Next, the another SPA is taken with $r_{0,2} < r_{0,1}$. The approximation procedure is repeated, and the second set of parameters A_2 and B_2 is obtained. Taking consequently SPAs along the signature and applying the LSM approximation procedure, the number of sets of A_i and B_i can be obtained. It should be noted that the MWS V_m, the RMW r_m, friction factor k, exponent n, and the Coriolis parameter f are the same for all approximation runs and different SPAs. Because of that, the values of the HLS coefficients pertained to different sets of evaluated data should satisfy the relationships:

$$A_i y_{0,i}^{n+1} = A_{y,i} = \frac{y_m^n}{k(n+1)r_{0,0}} V_m = A_y, \quad (32)$$

$$B_i = \frac{f}{k} = B, \quad (33)$$

where A_y and B are constants pertained to the given spiral element, $r_{0,0}$ is the polar radius of remotest SPA where the spiral signature can be described by the HLS, $y_m = r_m/r_{0,0}$, and $y_{0,i} = r_{0,i}/r_{0,0}$. Relationships (32) and (33) together combine the condition of the HLS stationarity. That is, if a spiral signature is really described by the HLS, then the parameters $A_{y,i}$ and B_i should be independent (theoretically) on selection of the SPA. Let us assume that the stationarity condition is satisfied within a polar radius range that corresponds to SPAs of indexes from i_s to i_e. The definition of the stationary part above is related to the general case, where a spiral signature is represented by N points with numbers $i = 1, 2... i_s ... i_e ... N$. The first SPA that is remotest from the center of the cyclone ($i = 1$) is selected arbitrarily, and a SPA with index i_s has polar radius $r_{0,0}$. This ordered nest of point coordinates, where the polar radii of the points are normalized by a distance from the remotest SPA, combines the normalized profile (NP) of a spiral signature. A logarithmic modification of NP, where relative radii are substituted by their natural logarithms, i.e., $\varphi_i = f(\ln y_i)$, is a logarithmic NP (LNP). The sequential approximation of a spiral signature is performed by the selection of consequent SPAs toward the cyclone center by a given number of points. The deficiencies of conditions (32) and (33) are primarily due to two reasons. The starting of a nonstationary part of a spiral signature $\{i = 1 \ldots (i_s\text{-}1)\}$ is due to the incorrect tracing of streamlines by clouds or the deviation of a streamline from the theoretical HLS for some reason. Regarding the "floor bounding" of the stationary part under SPA with $i \geq i_e + 1$, it is possible to show that this feature is due to the change of wind and friction regimes near the area of maximum winds. Beginning from this point, the increase of wind speed becomes slower than that of power n_{opt}. It might be assumed also that due to high wind speed, the friction factor becomes lower [21] or its dependence on wind speed becomes not linear. At that, the approximation estimates of the HLS coefficients were conducted under n_{opt} divergent from their stationary values, $\tilde{B}(n_{opt}) \rightarrow \infty$ and $\tilde{A}_y(n_{opt}) \rightarrow 0$. Therefore, the HLS coefficients should be selected from the results of the approximation of a spiral signature within a range where the stationary condition is satisfied. On the other hand, the distance from the center, where such divergence occurs, can be taken as an additional parameter of the cyclonic vortex (radius of divergence). Simulation experiments validated the stationarity and "divergent" property of the HLS coefficients.

3.1.3 *An example of HLS approximation with LSM*

As an example of the application of the above approach to the approximation of the SCRB by the HLS, we consider the estimate of the MWS from an image from the geostationary satellite GMS-5 in the visible range of TC Mitag, at 01:31 UTC on March 5, 2002 (**Figure 5**).

The corresponding logarithmic normalized profile of the annotated signature is shown in **Figure 6**.

As can be seen from **Figure 6**, the "regularity" of the profile is broken in the range lny = (− 0.6)–(− 0.7). Apparently this is due to not quite correct annotation of the final segment of the SCRB. It would have to be carried out along the steeper contrast border observed in this area. Therefore, the main part of the profile in the range lny = 0–(− 0.7) was subjected to approximation, only. The main characteristics of the signature were radius of the most distant starting point of the signature r_0 = 447 km, r_m = 130 km, and f = 3.3 10–5 s^{-1} (latitude 13.29°N). As the consecutive

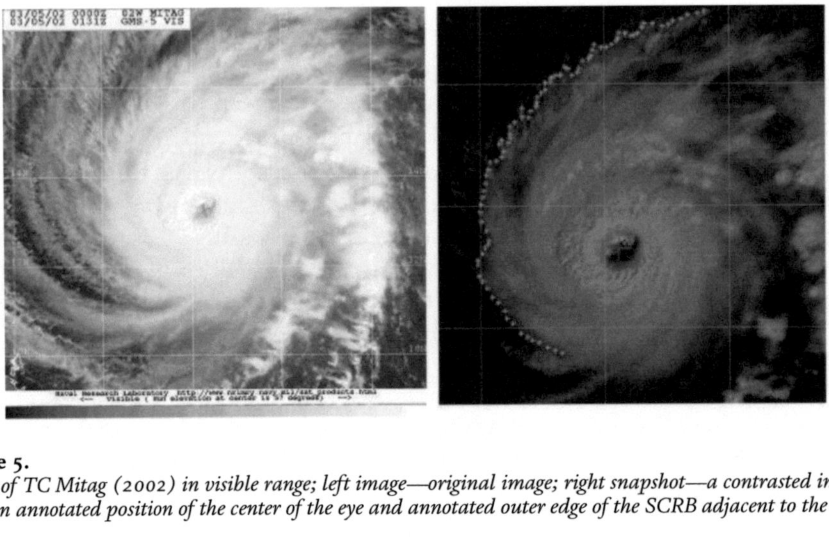

Figure 5.
Image of TC Mitag (2002) in visible range; left image—original image; right snapshot—a contrasted image with an annotated position of the center of the eye and annotated outer edge of the SCRB adjacent to the cyclone core.

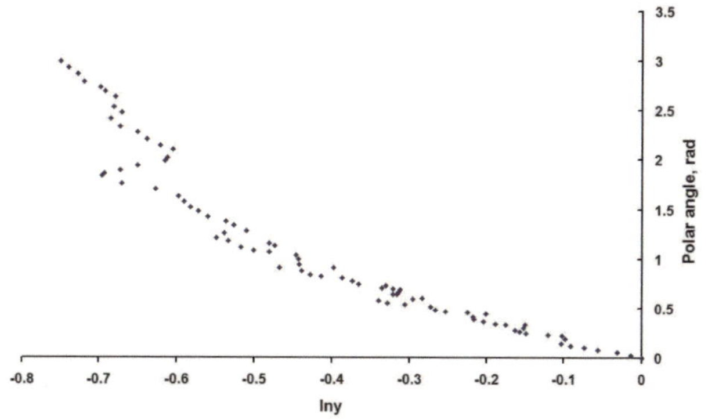

Figure 6.
Logarithmic normalized profile of SCRB annotated in ***Figure*** *5 (right snapshot).*

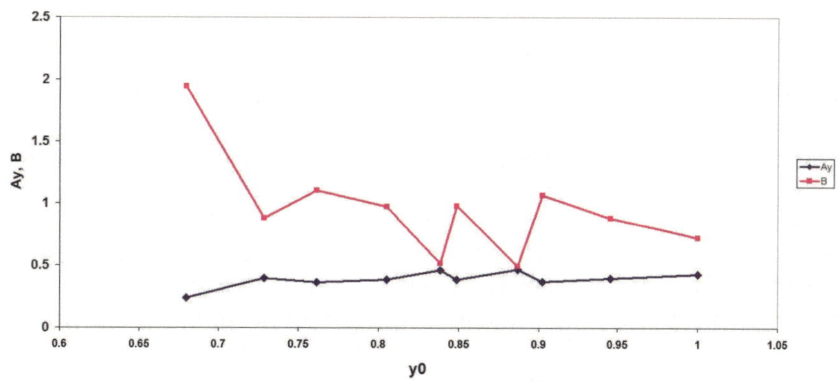

Figure 7.
Graph of stationarity of HLS approximation at n = 1.1.

tests showed, the stationarity conditions were best performed for both HLS coefficients in this particular case with the hyperbolic index $n = 1.1$ (**Figure 7**).

The average values of the coefficients A_y and B in the stationary segment were $A_y = 0.41 \pm 0.04$ and $B = 0.85 \pm 0.22$, which corresponds to $V_m = 58.1 \pm 16.1$ ms^{-1}. The meteorological assessment of the TC intensity for this point in time was 115 kts (59.2 ms^{-1}), which can be considered a satisfactory coincidence even taking into account the difference in altitudes for which the meteorological assessment is made (10 m) and the upper boundary of clouds for which the HLS estimate was calculated. According to experimental data, the MWS in a TC is observed at the level of 850–900 hPa: V_m(850 hPa) [22]. The average speed at the level of cloud top V_m(240 hPa) makes up about 70–80% of V_m(850 hPa) [22, 23]. The surface wind speed (at 10 m height) is smaller than V_m(850 hPa) by about 25–30% [24]. Therefore, considering that the configuration of the spiral signature depends on wind speed at the height of its existence, the HLS estimates can be compared (as a first guess) with surface data of meteorological services.

3.2 Assimilation technique

3.2.1 Motivation

As follows from the previous section, for the HLS approximation of SCRB by the LSM with the subsequent checking for stationarity, a lot of operations are needed to be performed. This requires considerable time and, therefore, is unlikely to have a prospect for use in operational practice. In addition, annotation of the spiral structure on the image is, to a certain extent, a subjective process as was shown by the example of **Figure 6**. Moreover, sometimes, as was shown in [11], the LSM approximation leads to nonphysical values of HLS coefficients due to incorrect annotation of the spiral structure. The main problem of the HLS approximation with the LSM is that, by its nature, this technique is applicable mainly to so-called thin or clearly depicted spiral structures, which have a small width or sharp contour in satellite or radar images, and are suitable for its uniquely annotation. However, the same thin spiraling bands very often turn out to be squall lines, which are not related to streamlines. These circumstances stimulated the search for another technique that would be applicable to typical SCRBs, having a noticeable width and fuzzy contours. This technique, called the assimilative technique, is discussed below.

3.2.2 A principle of assimilative technique

When choosing a technique for determining the HLS coefficients (3) and (4), it should be noted that a spiral rainband observable on a radar image has a finite width and is a mapping of the resultant involvement of cloud-rain particles in the region affected by the streamline. In this case, parameters of the HLS describing all possible streamlines within the rainband should be considered as "equally possible." In this technique, an approach based on the allocation of HLSs "fitting" into geometric boundaries of a rainband and determining the "expected" (mean) and "modal" HLSs was used. The fitting spirals were designated as "signatural" HLSs. The signatural HLSs have coefficients A (3) and B (4) which are determined using different combinations of V_m, k, and n. This triplet of parameters is hereafter referred to by the term "physical characteristics." The mean and modal values of a selected physical parameter over all signatural HLSs are taken as the corresponding statistical estimates of this parameter. The above procedure is called "assimilative HLS approximation." For the first time, this technique was used for processing of satellite infrared images of a TC in [13]. Further, it was improved in [14]. The primary changes in enhanced technique are the limitation of a range of possible maximum wind speeds resulted from the spiral band processing and finding the modal value of V_m distribution over all signatural HLSs in addition to the mean value. The technique was called "assimilative" due to a priori assignment of the type of the approximating function (HLS) and the range of variation of its basic parameters. The essentials of the technique are provided in [14].

4. Application of HLS assimilation technique for assessment of the maximum wind speed from satellite and radar data

4.1 HLS assimilation approximation of spiral rainbands applied to satellite IR images

4.1.1 Typhoon Phanfone (2002)

The first attempt to apply the HLS assimilation approximation to spiral rainbands on satellite IR images was undertaken by exploring data from the GMS-5 geostationary satellite during the monitoring of typhoon Phanfone existing in the Pacific Northwest in August 2002 [13]. The source of satellite data was the archive of IR images from the Naval Research Laboratory (NRL, USA; http://www.nrlmry.navy.mil/sat_products.html). Meteorological data were taken from the Navy/Air Force Joint Typhoon Warning Center (JTWC-WP; Hawaii, USA) and Regional

Specialized Meteorological Center (RSMC-JMA; Tokyo, Japan). IR spiral rainband patterns pertain to the top of cyclone's cloud system (280–230 hPa, altitude 10–12 km). The rationale for comparison of the MWS resulted from the HLS processing with the best track data is the same as mentioned above.

Twenty-one images were processed which corresponded to the period from 21:31 UTC on August 14, 2002, till 12:31 UTC on August 18, 2002. Their results were subsequently united to six estimates. The length of the period covered by the analysis was 84 h. The estimates of the MWS are used to identify the stages of intensification and weakening of TC which are satisfactorily synchronized with the respective data of the Navy/Air Force Joint Typhoon Warning Center (JTWC). The correlation with data of the Regional Specialized Meteorological Center (RSMC-JMA) is observed only for the stage of intensification. Data of JTWC did not correlate either with RSMC data at the stage of TC weakening. In terms of the absolute value of the MWS, JTWC data exceed RSMC data for the moment of maximum TC intensification. It should be noted that data of the mentioned meteorological services differ from each other by 33% in terms of maximum TC intensity. The estimates of the MWS based on the HLS approximation occupy intermediate position between the above data. A detailed description of this study is given in [13].

4.1.2 Hurricane Rita (2005)

In the current paper, an example of application of the HLS enhanced assimilation technique under processing IR images of a TC is provided exploring observation data of Hurricane Rita (AL182005) from geostationary satellite GOES-10 when the hurricane was in Gulf of Mexico. An image acquired at 6:30 UTC 23 September 2005 has been processed (**Figure 8**).

The image contains two spiral rainbands. They are the outer impressive band that starts from the northwest corner of the image and the inner one that starts from the north. Both bands reach the core of the cyclone close to its south sector. The radius of maximum wind has been assessed to be 70 km as a distance from the eye's

Figure 8.
IR image of Hurricane Rita 09/23/2005 at 06:30 UTC (eye center location: 26.58° N; 90.53° W) from GOES-10 geostationary satellite; source: Naval Research Laboratory (NRL, USA; http://www.nrlmry.navy.mil/sat_products.html); file name, 20050923_0630_goes10_x_ir1km_18LRITA_120kts-924mb-265 N-907 W; original geographic grid is applied at intervals of two degrees; color temperature scale (roughly), blue is "−40° C," green is "−60°C," and red is "−70°C."

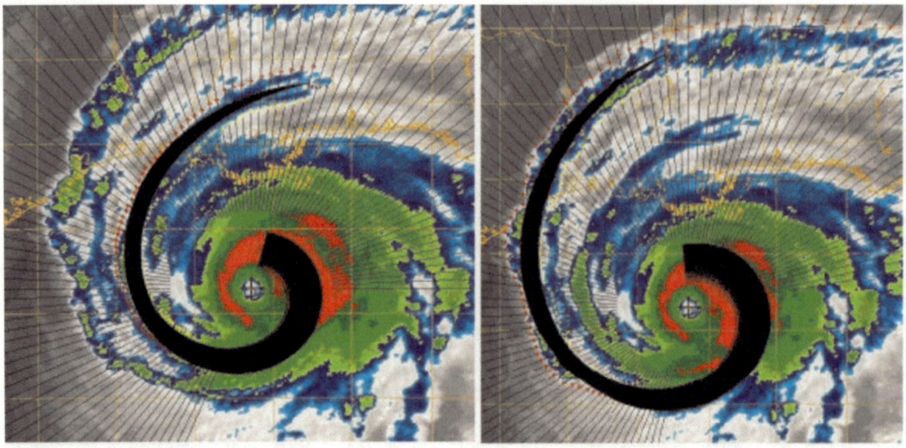

Figure 9.
*Illustration of HLS approximation (black spiral sector) of inner rainband (left picture) and the west sector of outer rainband (right picture) shown in **Figure 8**.*

Rainband	Maximum wind speed, kts	
	Modal MWS derived from HLS approximation	Best track*
Inner band	117.7 ± 8.7	115
Western sector of outer band	114.1 ± 13.6	

*Extrapolated from [25].

Table 1.
Maximum wind speed derived from HLS approximation in comparison with the best track data of hurricane Rita.

center to the middle of a convective eyewall. Illustrations of the HLS approximation of both rainbands are provided in **Figure 9**. Resulted maximum wind speeds in comparison with the best track data [25] are presented in **Table 1**.

Following the HLS technique feature that presumes to use for approximation a signature closest to the TC core (where a streamline impact on cloud organization is most pronounced), the western sector of the outer rainband was approximated only. As follows from **Table 1**, the approximation of both rainbands results in the modal maximum wind speeds close to the best track speed with the acceptable accuracies (less than 12%). It should be noted that among other things, this illustrative example also shows the possibility of increasing the reliability of the HLS approximation by combining multiband estimates. In particular, the combined weighted estimate of MWS based on the data provided in **Table 1** is 116.6 ± 7.4 kts.

4.2 HLS assimilation approximation of spiral rainbands applied to airborne and coastal radar images

4.2.1 Comparison of operational and HLS estimates of the intensity of TC Irma (AL112017) based on the airborne and the best track data

Maximum wind speeds in Hurricane Irma (AL112017) were estimated in [14] using its rainband radar signatures acquired by the NOAA Hurricane Research Division during routine aircraft missions into the hurricane. Most appropriate radar and other accompanied data were taken from the NOAA Hurricane Research Division (HRD) archive acquired for Hurricane Irma (AL112017) that existed in the

Atlantic basin from August 30 to September 12, 2017, and reached category 5 intensity. The radar data from airborne radars were taken from the HRD's Atlantic Oceanographic and Meteorological Laboratory (AOML) website (http://www.aoml.noaa.gov/hrd/data_sub/radar.html). The best track and aircraft data were taken from the National Hurricane Center's Tropical Cyclone Report [26]. The data pertain to one of eight aircraft missions to the cyclone. This mission was conducted from morning to afternoon September 5, 2017. The numerical outcomes of the HLS approximation are listed in **Table 2**.

The data of the first two time points were acquired at altitude from 5 km to 6 km, the six others from 2 km to 3 km. At the higher altitude, the MWS was approximately from 80 kts to 100 kts; at the lower level, the mean modal wind speed was about 161 $\pm$ 5 kts. This estimate is for a middle time point at approximately 10:45 UTC September 5 and manifests the maximum TC intensity that was estimated by the HLS approximation of all period of the HLS application from 21:16 UTC September 3 till 23:51 UTC September 8. As follows from [26], the wind speed of 164 kts measured directly by the aircraft at the flight level occurred approximately 10 h later. The best track MWS, which is for 10 m altitude, was about 150 kts. As per the best track data, the maximum intensity was 154 kts from noon September 5 to approximately 18:00 UTC September 6. The HLS estimates for this period are provided in [14] and amounted to 158–146 kts. These results indicate the satisfactory agreement of the HLS approach with in situ data (compare 161 kts and 164 kts). Lower speeds evaluated at high levels (two first time points in **Table 2**) follow the contemporary understanding of the vertical tangential wind profile in a TC (e.g., [24]) that presumes the decreasing of the wind speed up from the level of the maximum wind at altitude approximately 1–1.3 km (850–900 mb). The average error of the HLS approximation for the entire observation time ($\sim$ 130 h) of the comparative analysis with the best track data given in [14] was no more than 5%.

4.2.2 The maximum wind speed in TC Irma (2017) by the HLS approximation of the rainband signatures from the coastal San Juan radar in comparison with best track data

During the passage of the TC Irma near the island of Puerto Rico, the cyclone was in the survey zone of the weather radar WSR-88D installed in the city of San Juan. The base for processing was the reflectivity image at 21:15 UTC 6 September

Time point no.	Altitude, m	UTC	Observation time, hours	V_{m_mdl}		Error V_{m_mdl}		R_m, km
				m s^{-1}	kts	m s^{-1}	kts	
22	6089	8:41:44	56.68	53	103.0	2.0	3.9	30
23	5244	8:47:58	56.78	40	77.8	3.0	5.8	26
24	2999	9:27:39	57.45	75.5	146.8	2.5	4.9	34
25a	2999	9:43:20	57.72	76	147.7	4	7.8	33
25b	3000	9:48:04	57.80	89	173.0	1	1.9	30
26	2510	10:31:26	58.52	79	153.6	3	5.8	34
27	2369	11:01:38	59.02	89	173.0	7	13.6	30
28	2531	12:01:32	60.02	88	171.4	4	7.8	31

Table 2.
Results of the HLS approximation of radar spiral signatures of TC Irma acquired during airborne sounding on September 5, 2017 (collection index 20170905H1).

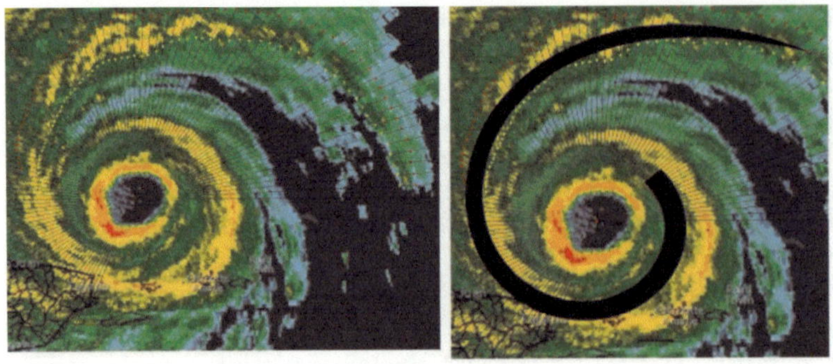

Figure 10.
Left image—reflectivity image of TC Irma from San Juan WSR-88D (2115 UTC September 6, 2017 [26]); right image—illustration of HLS approximation of the NW rainband.

2017 that is shown in **Figure 10** (left image) when the TC center was located at 18.9° N/65.4° W [25]. The image refers to the beginning of the cyclone weakening stage and has the double eyewall structure that indicates a double wind maximum. Under the HLS approximation, the RMW was selected within the outer eyewall, as follows from findings in [27, 28], and estimated to be about 62 km.

The modal and mean maximum wind speeds estimated from the HLS approximation of the northwest rainband (**Figure 10**, right image) were 74 m s^{-1} (143.9 kts) and 75.4 m s^{-1} (146.6 kts), respectively. Accordingly, these speeds are approximately by 6 kts and 3 kts lower than the best track speed 150 kts at this time.

5. Summary

A new approach to use the characteristics of spiral cloud-rain bands of a tropical cyclone, observed by ground-based and aircraft radars, as well as satellites in visible and infrared wavelengths, is considered. The physical substantiation of the proposed approach is (1) the assumption about the orientation of the SCRBs mainly along the streamlines and (2) the analytically derived streamline equation in the form of the hyperbolic-logarithmic spiral. It is shown that the logarithmic spiral usually used to describe the configuration of SCRB is only a special case of the HLS. Unlike the empirically applying logarithmic spiral, the HLS coefficients depend on the MWS and the friction factor. The analysis of changes in the configuration of the HLS depending on the intensity of a TC is conducted. An explanation for experimentally observed phenomenon of the "rounding" of the SCRB (i.e., a decrease in the crossing angle) with increasing intensity of the TC as a whole, as well as with approaching to the radius of maximum wind, is proposed. The similarity of the SCRB configuration to the logarithmic spiral in some cases is interpreted also. The technique for approximation of a SCRB in the form of the HLS based on the least squares method and on the assimilation procedure was developed for obtaining MWS estimates at the height of the SCRB location. Testing of the proposed approach was performed based on literature data from ground-based coastal and aircraft radars, data of regular aircraft reconnaissance missions in the TC, and satellite data available via the Internet. A certain disadvantage of the method is its applicability, as a rule, for mature tropical cyclones, where its cloudy field manifests well-defined SCRBs and a clearly defined circulation center (eye center). On the other hand, the physically based configuration of a SCRB as the hyperbolic-logarithmic spiral allows one to develop the method for estimating the position of the circulation center with the eye covered with clouds, examples of which are given in [11, 13]. The results obtained suggest that the development and improvement of the proposed approach will make it possible to use the radar and satellite information more fully to assess the physical characteristics of a TC. The HLS approach to retrieve the TC's intensity is particularly benefited for ground-based coastal radar probing of a TC before its landfall and the absence of aircraft reconnaissance missions.

Author details

Boris Yurchak
Environmental Protection Agency, Washington D.C., USA

*Address all correspondence to: yurchak.boris@epa.gov

References

[1] Dvorak VF. Tropical cyclone intensity analysis and forecasting from satellite imagery. Monthly Weather Review. 1975;**103**:420-430

[2] Olander TL, Velden CS. The advanced Dvorak technique: Continued development of objective scheme to estimate tropical cyclone intensity using geostationary infrared satellite imagery. Weather and Forecasting. 2007;**22**: 287-298

[3] Wexler H. Structure of hurricanes as determined by radar. Annals of the New York Academy of Sciences. 1947;**48**: 821-844

[4] Senn HV, Hiser HW, Bourret RC. Studies of Hurricane Spiral Bands as Observed on Radar. National Hurricane Research Project. U.S. Department of Commerce. Report No. 12; 1957. 13 p

[5] Fung IY. The organization of spiral rain bands in a hurricane. Doctoral dissertation thesis. Massachusetts: Massachusetts Institute of Technology; 1977. 140 p

[6] Fernandez W. Organization and motion of the spiral rainbands in hurricanes: A review. Ciencia y Tecnología. 1982;**6**(1–2):49-98

[7] Houze RA Jr. Clouds in tropical cyclones. Review. Monthly Weather Review. 2010;**138**:293-344

[8] Wang Y. Recent research progress on tropical cyclone structure and intensity. Tropical Cyclone Research and Review. 2012;**1**(2):254-275

[9] Lahiri A. A study of cloud spirals of tropical cyclones. Mausam. 1981;**32**(2): 155-158

[10] Burton A, Velden C. Curved band patterns. In: Proceedings of the International Workshop on Satellite Analysis of Tropical Cyclones. Report Number TCP-52. Honolulu, Hawaii, USA: World Meteorological Organization; 2011, 2011. pp. 65-67

[11] Yurchak BS. Description of cloud-rain bands in a tropical cyclone by a hyperbolic-logarithmic spiral. Russian Meteorology and Hydrology. 2007; **32**(1):8-18

[12] Yurchak BS. Formula for spiral cloud-rain bands of a tropical cyclone. In: Proceedings of the 28th Conference on Hurricanes and Tropical Meteorology. Orlando, FL; 2008, 2008. 5 p. Available online: http://ams.confex.com/ams/28Hurricanes/techprogram/programexpanded_471.htm

[13] Yurchak BS. Estimation of tropical cyclone intensity from the satellite infrared images of its spiral cloud bands. Russian Meteorology and Hydrology. 2018;**43**(9):581-590

[14] Yurchak BS. An estimate of the hurricane's intensity from radar data using hyperbolic-logarithmic approximation. International Journal of Remote Sensing. 2019. DOI: 10.1080/01431161.2019.1635288

[15] Willoughby HE, Marks FD, Feinberg RJ. Stationary and moving convective bands in hurricanes. Journal of the Atmospheric Sciences. 1984; **41**(22):3189-3211

[16] Willoughby HE. The dynamics of the tropical cyclone core. Australian Meteorological Magazine. 1988;**36**:183-191

[17] Batchelor GK. An Introduction to Fluid Dynamics. Cambridge: Cambridge University Press; 1967. 634 p

[18] Guralnik II, Dubinskii GP, Mamikonova SV. The Meteorology. Handbook for Colleges.

Gidrometeoizdat, Leningrad, 1972. 416 p.[in Russian]

[19] Wendell HE. Meteorology-Theoretical and Applied. Kennelly Press. (Copyright, 1944 by Hewson EW and Longley RW); 2007. 468 p

[20] Anthes RA. Tropical cyclones. Their evolution, structure and effects. AMS, Meteorological Monographs. 1982; **19**(41):210

[21] Powell MD, Vickery PJ, Reinhold TA. Reduced drag coefficient for high wind speeds in tropical cyclones. Nature. 2003;**422**:279-283

[22] Frank WM. The structure and energetics of the tropical cyclone I. storm structure. Monthly Weather Review. 1977;**105**:1119-1135

[23] Franklin JL, Lord SJ, Feuer SE, Marks FD Jr. The kinematic structure of hurricane gloria (1985) determined from nested analysis of dropwindsonde and doppler radar data. Monthly Weather Review. 1993;**121**:2433-2450

[24] Franklin JL, Black ML, Valde K. GPS dropwindsonde wind profiles in hurricanes and their operational implications. Weather and Forecasting. 2003;**18**:32-44

[25] Knabb KD, Brown DP, Rhome JR. Tropical Cyclone Report Hurricane Rita, 18–26 September 2005. National Hurricane Center; 2006. 36 p. Available online: https://www.nhc.noaa.gov/data/tcr/AL182005_Rita.pdf

[26] Cangialosi JP, Latto AS, Berg R. National Hurricane Center Tropical Cyclone Report. Hurricane Irma (AL112017). 30 August-12 September 2017; 2018. 111 p. http://www.aoml.noaa.gov/hrd/Storm_pages/irma2017/index.html

[27] Willoughby HE, Clos JA, Shoreibah MG. Concentric eye walls, secondary wind maxima, and the evolution of the hurricane vortex. Journal of the Atmospheric Sciences. 1982;**39**(2):395-411

[28] Samsury CE, Zipser EJ. Secondary wind maxima in hurricanes: Airflow and relationship to rainbands. Monthly Weather Review. 1995;**123**:3502-3517

Section 3

Remote Sensing and Modeling

Chapter 6

NASA Global Satellite and Model Data Products and Services for Tropical Cyclone Research

Zhong Liu, David Meyer, Chung-Lin Shie and Angela Li

Abstract

The lack of observations over vast tropical oceans is a major challenge for tropical cyclone research. Satellite observations and model reanalysis data play an important role in filling these gaps. Established in the mid-1980s, the Goddard Earth Sciences Data and Information Services Center (GES DISC), as one of the 12 NASA data centers, archives and distributes data from several Earth science disciplines such as precipitation, atmospheric dynamics, atmospheric composition, and hydrology, including well-known NASA satellite missions (e.g., TRMM, GPM) and model assimilation projects (MERRA-2). Acquiring datasets suitable for tropical cyclone research in a large data archive is a challenge for many, especially for those who are not familiar with satellite or model data. Over the years, the GES DISC has developed user-friendly data services. For example, Giovanni is an online visualization and analysis tool, allowing users to visualize and analyze over 2000 satellite- and model-based variables with a Web browser, without downloading data and software. In this chapter, we will describe data and services at the GES DISC with emphasis on tropical cyclone research. We will also present two case studies and discuss future plans.

Keywords: satellites, models, data, data services, NASA

1. Introduction

Tropical cyclones form over vast tropical oceans where in situ observations are sparse and discontinuous. The lack of observational data over these areas historically has been a major obstacle for tropical cyclone research and other weather and climate-related studies. Understanding the complex atmospheric and oceanic processes, and their interactions at multiple scales (e.g., convective, synoptic) over the life cycle of tropical cyclones, requires multiscale, multi-platform observational networks. It has been a great challenge to design, deploy, and maintain such networks without interruption, particularly in the harsh environments imposed by these extreme phenomena. Since the satellite era began, data collected from satellites, along with model reanalysis data, have played an important role in providing continuous global observations, filling in data gaps, and enabling research on weather and climate on different scales.

The concept of using satellites to observe Earth's weather and climate was developed as early as 1946 [1]. NASA launched the first successful, weather satellite,

TIROS-1 (Television InfraRed Observational Satellite) on April 1, 1960 [1, 2]. In 1964, the Nimbus project was initiated and a total of 7 experimental meteorological satellites were launched over a 14-year time period (1964–1978) [1, 2]. Since then, weather and climate research have come to rely heavily on long-term, consistent satellite observations from multiple operational space-borne platforms to continuously observe the Earth's atmospheric and surface conditions.

The Earth-observing satellite era began in earnest after NASA transferred the technology to the National Oceanic and Atmospheric Administration (NOAA) in the 1970s. This was followed by several operational weather satellites series, including the Polar-orbiting Operational Environmental Satellites (POES) and the Geostationary Operational Environmental Satellites (GOES), to provide continuous global weather observations. Meanwhile, the Defense Meteorological Satellite Program (DMSP), also launched in the 1970s, provided additional observations of global weather events. These series evolved into the constellation of weather satellites operating today, using research and operational satellites from domestic and international organizations to provide the frequent, global observations necessary for improved understanding and forecasting of Earth's complex weather systems.

Atmospheric 3-D winds, air and sea surface temperatures, pressure, precipitation, water vapor, aerosols, etc. are among the fundamental variables for tropical cyclone research and applications. As aforementioned, few in situ observations are available over vast and remote tropical oceans. Direct measurements of these variables are difficult both from surface and space. Over the years, satellite-based algorithms have been developed and improved to derive these essential variables from few key measurements such as radiances observed onboard satellites, and their datasets are archived and distributed to support tropical cyclone research.

Established in the mid-1980s (**Table 1**), the Goddard Earth Sciences Data and Information Services Center (GES DISC), as one of the 12 NASA Distributed Active Archive Centers (DAACs), archives and distributes satellite and model data for a range of Earth science disciplines [3], such as precipitation, atmospheric dynamics, atmospheric composition, and hydrology, derived from well-known NASA Earth's satellite missions (e.g., the Tropical Rainfall Measuring Mission (TRMM), the Global Precipitation Measurement (GPM)) as well as model assimilation projects (MERRA-2, NLDAS). These data have been widely used in tropical cyclone research.

To facilitate data access, the GES DISC has developed user-friendly data services for researchers around the world (**Figure 1**). For example, the Geospatial

• Mid-1980s—one of two original DAACs (with NASA's Langley Research Center)—"Goddard DAAC"
• 1990s Version 0 era
• AVHRR (Advanced Very High-Resolution Radiometer) pathfinder
• TOVS (TIROS Operational Vertical Sounder) pathfinder
• SeaWIFS (Sea-viewing Wide Field-of-view Sensor)
• UARS (Upper Atmosphere Research Satellite)
• 1997—TRMM (first EOS launch)
• 2000/2002—Terra/Aqua MODIS (Moderate-Resolution Imaging Spectroradiometer)
• 2005/2006—EOSDIS evolution—split Goddard DAAC:
• GES DISC—atmosphere/hydrology/climate focus
• Level-1 and Atmospheres Archive Distribution System (LAADS)—MODIS instrument focus
• Ocean Biology DAAC—ocean biology focus

Table 1.
A brief history of the GES DISC.

Interactive Online Visualization ANd aNalysis Infrastructure (Giovanni) [4–7] is a powerful online visualization and analysis tool, allowing users to visualize and analyze over 2000 satellite- and model-based variables with a Web browser, without downloading data or software. Locating datasets suitable for tropical cyclone research in a large data archive is a challenge for many users, especially those who are not familiar with satellite data. For example, a search for "precipitation" in the GES DISC Web site [3] can return over 400 results. To facilitate data access, the GES DISC has recently developed a "data list" (also known as "variable set") concept that groups relevant variables from different datasets together to serve specific research needs. A prototype data list targeting hurricane study has been implemented.

The chapter is organized as follows: first, we give a brief overview of NASA satellite and model data at GES DISC, followed by introducing datasets for tropical cyclone studies, data services, case studies, and summary with future directions.

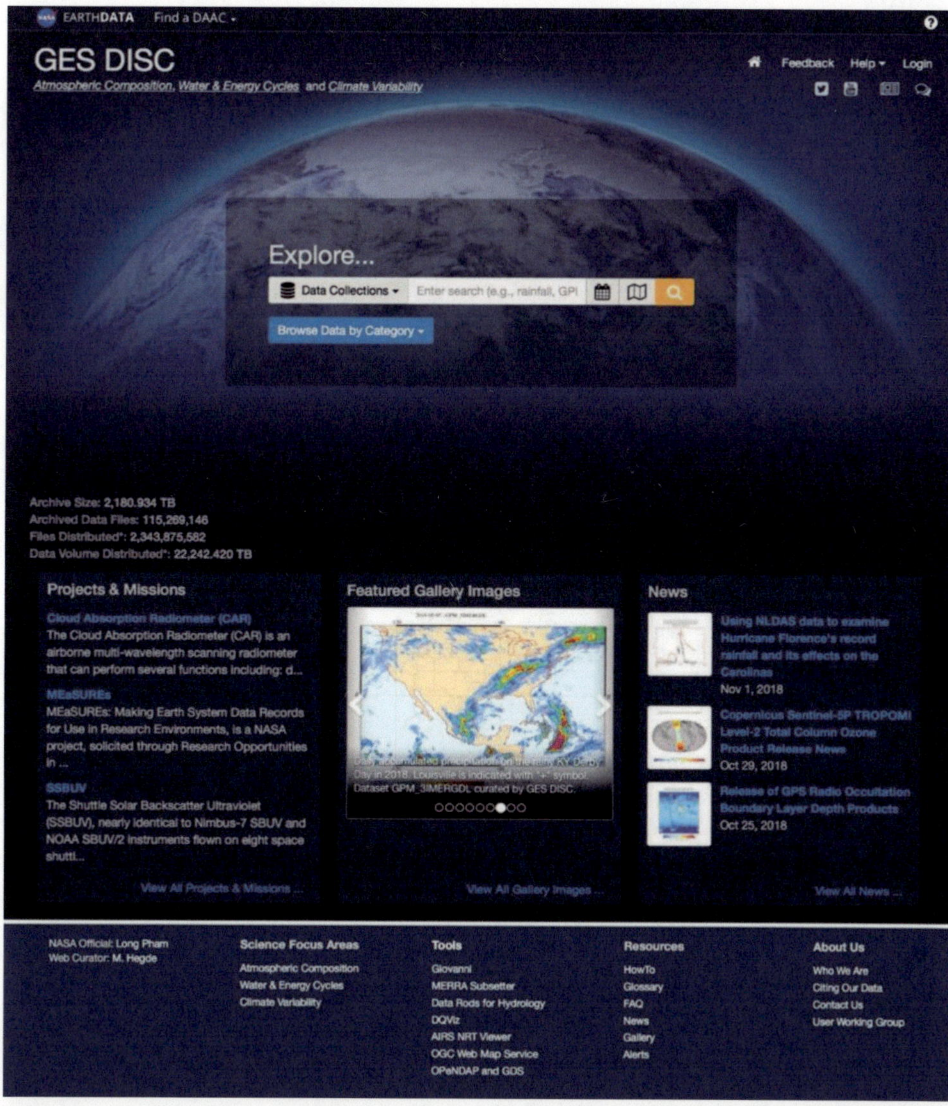

Figure 1.
The GES DISC Web site. This all-in-one design allows search for dataset and information at GES DISC. Users can access the latest news, projects, missions, tools, resources, and more in this Web site.

2. Overview of NASA satellite mission and model data collections at GES DISC

The GES DISC archives and distributes data from a range of satellite observations, models, ground measurements, and field campaigns in multiple Earth science disciplines including global precipitation, atmospheric dynamics, hydrology, and atmospheric composition with a total volume of 2.3+ Petabytes consisting of 100+ million data files covering 3000+ public and restricted collections. Over 1200 data collections are being curated at GES DISC. **Table 2** lists their satellite missions

Atmospheric composition missions:

- Nimbus 1–7* BUV, SBUV, TOMS
- Shuttle SBUV*
- UARS*
- Aqua AIRS
- Aura HIRDLS*, OMI, MLS
- ACOS*
- SNPP Sounder, OMPS
- JPSS-1 Sounder, OMPS
- OCO-2
- Copernicus Sentinel 5P
- TOVS Pathfinder*

Water cycle/precipitation missions:

- TRMM*
- GPM
- SMERGE

Climate variability/solar missions:

- SORCE
- TCTE
- TSIS
- CAR

Future assigned missions:

- OCO-3
- TROPICS
- Copernicus Sentinel 6
- GeoCarb

Model projects:

- MERRA*/MERRA-2
- NLDAS, GLDAS, FLDAS, NCA-LDAS

Other projects:

- MEaSUREs: Making Earth System Data Records for Use in Research Environments
- CMS

**End-of-mission/project.*

Table 2.
Past, current, and future NASA satellite missions that are associated with their data products curated at GES DISC.

Data services and support at GES DISC
Metadata support, documentation, metrics:
• Assignment of DOIs
• Includes recommended dataset citation, hosting of dataset landing pages, documentation
• Generation of metadata records, publication to the EOSDIS Common Metadata Repository (CMR)
• Publication of data **distribution metrics** to the EOSDIS Metrics System (EMS)
Web-based discovery and access to products (value-added services on data):
• Giovanni
• Subsetting, reformatting and regridding
• Access protocols (e.g., OPeNDAP)
User services—provide tiered support in data access and use:
• GES DISC User Services (first tier)
• GES DISC science data specialist (second tier)
• Collaboration with science team subject matter experts (third tier)
Community Engagement:
• Workshops and webinars on the use of data and relevant services
• Conference participation, publications, news releases
• Engagement with Applications Community
• Applied Remote Sensing Training Group (ARSET), Disasters Working Group, Heath and Air Quality Applied Sciences Team (HAQAST), Land and Atmospheres near-real-time Capabilities for EOS (LANCE).

Table 3.
A list of data services and support at GES DISC.

including the past, current, and future satellite missions. **Table 3** lists basic data services and user support. More details about data services are described in Section 4.

The GES DISC is a certified trusted repository of Earth science data. Increasingly, funding organizations and publishers require data to be published to certified data repositories adhering to FAIR principles—(Findability, Accessibility, Interoperability, and Reusability). The GES DISC is a regular member of the International Council for Science (ICSU) World Data System (WDS). Established to archive and distribute data from the 1957–1958 International Geophysical Year, WDS spans a range of scientific disciplines data at 52 centers in 12 countries who adhere to the established principles. The GES DISC is also registered as a scientific data repository through re3data.org and meets the repository criteria including DOI assignments, dataset landing pages, dataset documentation, redundant archive (backups), data integrity checks, and user services. This registry is used by high-impact journals such as Nature Scientific Data [8].

3. Datasets for tropical cyclone research

There are many research areas associated with tropical cyclones such as cyclone genesis, intensification, track forecasting, rainfall amounts, etc. While GES DISC archives many variables required to conduct such research, other variables may be located at other NASA DAACs (such as sea surface state and temperature). It can be challenging enough to find relevant variables for a specific research area from a large collection of satellite observations from a single data archive—locating relevant datasets from multiple data centers is even more challenging. Due to the page limit, in this section, we can only present a brief overview of several key data collections at GES DISC for tropical cyclone research.

3.1 Brightness temperature collection

Brightness temperature, derived from radiance measured from satellite instruments, is a fundamental physical variable in satellite meteorology. Infrared satellite images have been available since the Nimbus era to support weather analysis and forecast. For example, infrared images are used in tropical cyclone monitoring and forecast operations at the NOAA National Hurricane Center and the U.S. Navy Joint Typhoon Warning Center. Animations made from infrared images are frequently used in daily local TV weather news, online weather news, and scientific presentations. The GES DISC archives brightness temperatures from infrared instruments from the Nimbus era up to more recent and current passive microwave satellite instruments from domestic and international research and operational satellites.

Datasets from the Nimbus data rescue project [9] consists of digitized black-and-white film images (**Figure 2**) and radiance data obtained by the Nimbus satellites during the 1960s, 70s, and 80s [10]. Related instruments onboard the Nimbus satellites

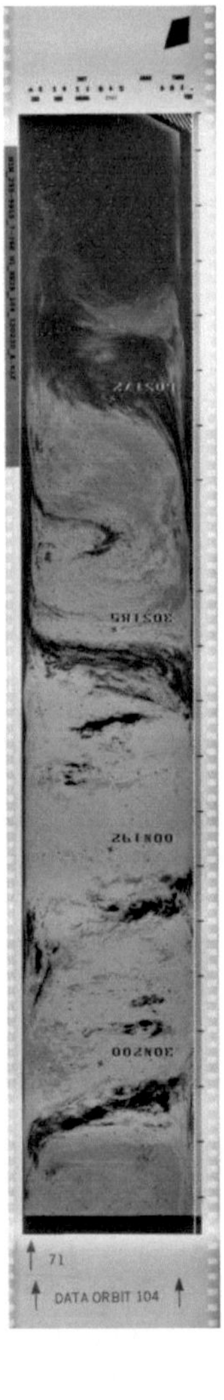

Figure 2.
A sample of HRIR/Nimbus-1 images of nighttime brightness temperature on 70 mm film.

over this time period are listed in **Table 4**. Negatives of photo facsimile 70-mm file strips were scanned and saved as JPEG 2000 digital files. There are 20 datasets from the Nimbus satellites (1–7) beginning on Aug. 28, 1964 and ending on May 9, 1985.

Tropical cyclone research requires frequent and continuous observations of the Earth's atmosphere to analyze and understand event development and processes. Operational geostationary satellites make such observations from cloud tops possible, although it is still challenging to continuously observe changes inside a weather system. The first geostationary satellite in operation is GOES 1, which was launched on October 16, 1975 [1]. As more operational geostationary satellites were added by different international agencies, infrared data from these satellites can be stitched and provide a near-global (60° N-S) coverage of the Earth's atmosphere [11, 12].

With support from the NASA Global Precipitation Climatology Project (GPCP) and by the Tropical Rainfall Measuring Mission (TRMM), the NOAA National Weather Service (NWS) Climate Prediction Center (CPC) has developed a globally merged (60° N-S) pixel-resolution IR brightness temperature dataset (equivalent blackbody temperatures), merged from all available domestic and international geostationary satellites [11, 12]. This half-hourly and 4 km x 4 km resolution dataset is also called the merged IR and is available at the GES DISC from 2000 onward [12]. **Figure 3** is a sample of the dataset, showing two tropical cyclones (Cilida and Kenanga) on December 20, 2018. In addition to tropical cyclone and other research, the merged IR dataset has been an important input for a number of algorithms that derive near-global IR-based precipitation estimates in several well-known satellite-based global precipitation products [13, 14] such as the Integrated Multi-satEllite Retrievals for GPM or IMERG [13].

Currently in operation, the GPM Microwave Imager (GMI) [15] is used as the reference standard to generate Level-1C common calibrated brightness temperature products from the GPM constellation consisting of both domestic and international satellites, based on the algorithms developed by the GPM intercalibration (X-CAL) working group [16]. These Level-1C products are transformed from their equivalent Level-1B radiance data. There are many applications of these passive microwave brightness products. For example, the GPM profiling algorithm (GPROF) uses these Level-1C products to generate hydrometeor profiles and surface precipitation estimates (**Figure 4**) used as input data in IMERG.

3.2 TRMM and GPM precipitation dataset collection

Launched in November 1997, the TRMM satellite (40° N-S), a joint mission between NASA and JAXA (the Japan Aerospace Exploration Agency), carried

- The High-Resolution Infrared Radiometer (HRIR) (Numbus-1, 2, 3)
- The Medium-Resolution Infrared Radiometer (MRIR) (Nimbus-3)
- The Satellite Infrared Spectrometer (SIRS) (Nimbus-3)
- The Nimbus-4 Selective Chopper Radiometer (SCR) (Nimbus-4, 5)
- The Infrared Interferometer Spectrometer (IRIS) (Nimbus-4)
- The Temperature-Humidity Infrared Radiometer (THIR) (Nimbus-4, 5, 6, 7)
- The Satellite Infrared Spectrometer (SIRS) (Nimbus-4)
- The Electrically Scanning Microwave Radiometer (ESMR) (Nimbus-5)
- The High-Resolution Infrared Radiometer (HIRS) (Nimbus-6)

Table 4.
Instruments onboard the Nimbus satellites.

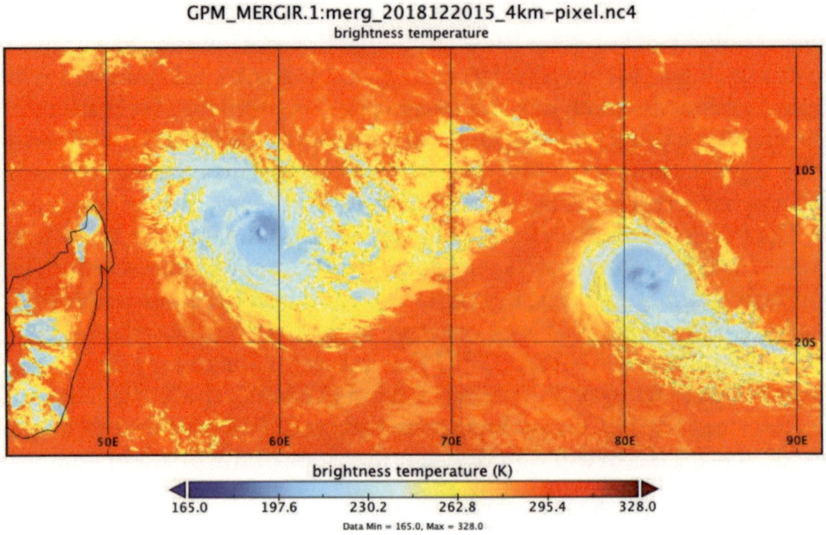

Figure 3.
Two tropical cyclones (Cilida on the left and Kenanga on the right) are seen from the NCEP/CPC merged IR dataset on December 20, 2018. The map was generated with the NASA GISS Panoply.

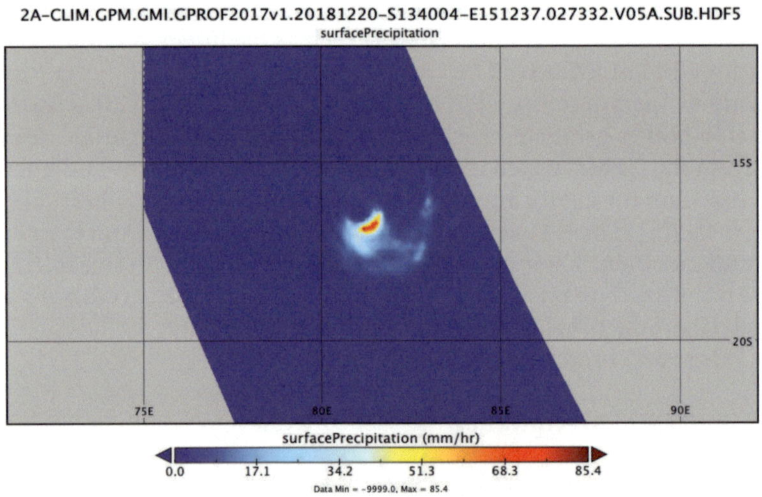

Figure 4.
GPM GMI surface precipitation from tropical cyclone Kenanga over the Indian Ocean on December 20, 2018. The data were generated by the GES DISC Level-2 subsetter and the map created with NASA GISS Panoply.

several precipitation-related instruments, including the first spaced-borne Ku-band precipitation radar (PR), a passive TRMM microwave imager (TMI), a visible and infrared scanner (VIRS), and a lightning imaging sensor (LIS) [17]. TRMM ended in April 2015. Over a 17-year period, TRMM provided observations that are used to produce groundbreaking 3-D images of rain and storms over vast and remote tropical oceans and continents. TRMM provides opportunities for researchers to understand characteristics of atmospheric systems through instantaneous measurements in different wavelengths from the onboard instruments.

TRMM data available at GES DISC [18] are listed in **Table 5**. They can be categorized in different processing levels, ranging from Level 1 to 3 [19]. Level-1 TRMM datasets consist of reconstructed and unprocessed instrument data at full-resolution data at Level-1, 1A, 1B, and 1C [19] from the three TRMM instruments. Level-2 TRMM datasets are derived geophysical variables at the same resolution and location as Level-1 source data such as the GPROF hydrometeor profiles and surface precipitation estimates. Level-3 TRMM datasets are Level-2 variables that are mapped on uniform space-time grid scales, ranging from 3 hourly to monthly. For example, the TRMM Multi-satellite Precipitation Analysis (TMPA) datasets [20–22] are Level-3 products, including both near-real-time and research grade products. The TMPA datasets have been widely used in research and applications around the world, especially in gauge sparse regions. TRMM products processed with GPM algorithms are also available [3]. Their data format and naming convections are consistent with those of GPM [3]. TRMM LIS data are archived at the Global Hydrology Resource Center (GHRC) [23]. Studies to investigate the relationship between lightning and precipitation have been reported (e.g., [24]).

Built on the success of TRMM, GPM [15] is another joint mission between NASA and JAXA to continue key measurements after the TRMM era. The main concept of GPM is to form an international constellation of research and operational satellites and use GPM as a core satellite that carries advanced radar and passive microwave radiometer instruments to measure precipitation from space as well as serve as a reference standard to unify precipitation measurements from other domestic and international satellites in the constellation [15].

Processing Level	**Dataset Name**	**Resolution**
Level-1	• 1B01: Visible and infrared radiance • 1B11: Passive microwave brightness temperature • 1B21: Precipitation radar power • 1C21: Precipitation radar reflectivity	5 ×5 km—16 orbits per day
Level-2	• 2A12: TMI hydrometeor profile • 2A21: Precipitation radar surface cross section • 2A23: Precipitation radar rain characteristics • 2A25: Precipitation radar rainfall rate and profile • 2B31: Combined rainfall profile (PR, TMI)	5×5 km—16 orbits per day
Level-3	• 3A11: Oceanic rainfall • 3A12: Mean 2A12, profile and surface rainfall • 3A25: Spaceborne radar rainfall • 3A26: Surface rain total • 3A46: SSM/I rain • 3B31: Combined rainfall • 3B42RT: 3-hour real-time TRMM Multi-satellite Precipitation Analysis (TMPA) • 3B42RT daily: 3B42RT derived daily product • 3B42: Research version of TMPA • 3B42 daily: 3B42 derived daily product • 3B43: Multi-satellite precipitation	• 3A11: 5°, monthly • 3A12: 0.5°, monthly • 3A25: 0.5°, 5°, monthly • 3A26: 5°, monthly • 3A46: 1°, monthly • 3B31: 5°, monthly • 3B42RT: 0.25°, 3 hourly • 3B42RT daily: 0.25°, daily • 3B42: 0.25°, 3 hourly • 3B42 daily: 0.25°, daily • 3B43: 0.25°, monthly

Table 5.
A list of TRMM datasets at GES DISC. TRMM products processed with GPM algorithms are also available [3]. Their data format and naming convections are consistent with those of GPM. More information is available in each dataset landing page.

In addition to the passive GPM microwave imager or GMI, a dual-frequency precipitation radar (DPR) has been added in GPM (**Figure 5a**). A new frequency (Ka-band) in the DPR is capable to detect light rain from space, which is one of challenges in satellite precipitation retrieval algorithms. The GMI carries four additional high frequency channels for measuring falling snow, compared to the TMI. The GMI's spatial resolution is improved significantly with a 1.2 m diameter antenna [15].

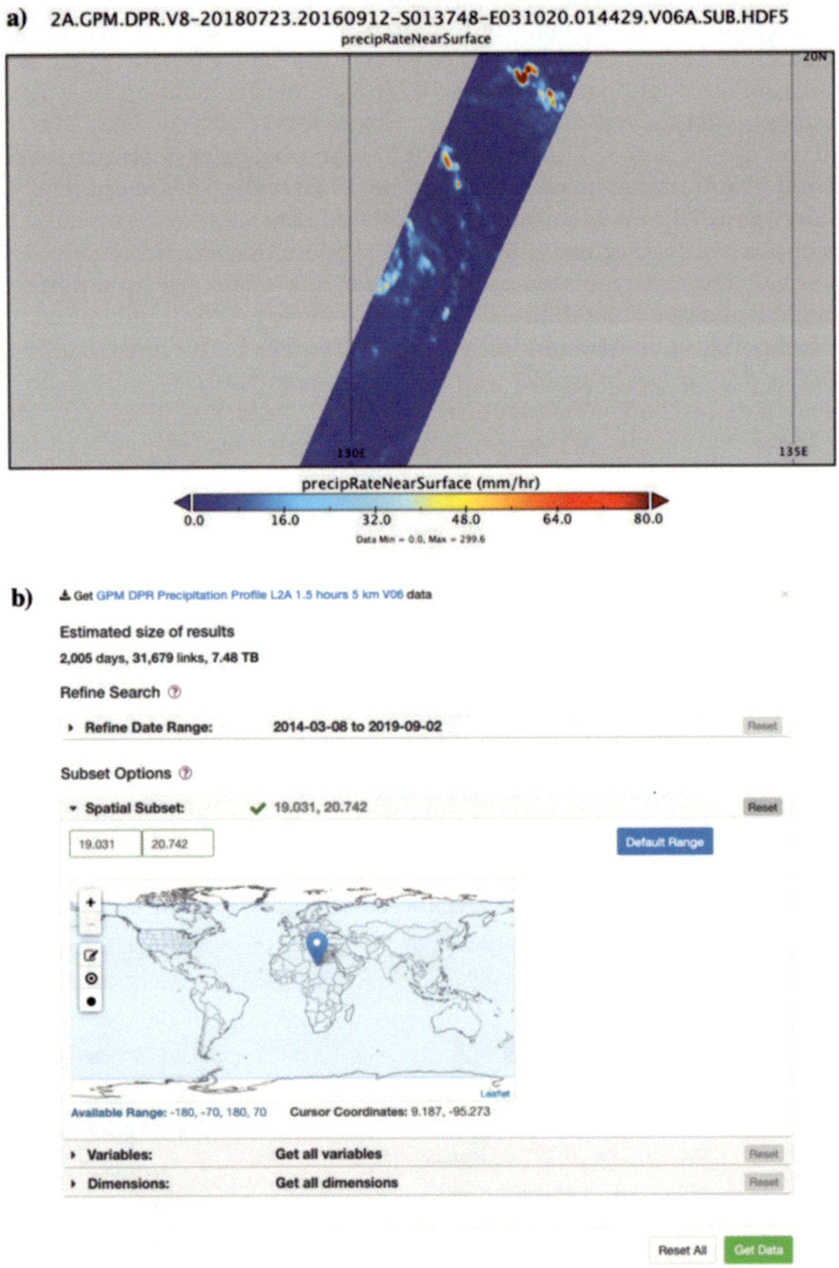

Figure 5.
(a): Near surface precipitation from the GPM DPR matched scans (MS), showing super typhoon Meranti on September 12, 2016 before impacting the Philippines, Taiwan, and Fujian Province. The data were generated with the Level-2 subsetter and the map with NASA GISS Panoply. (b): Three spatial subsetting options (box, circle, and point) in the Level-2 subsetter.

Processing Level	Dataset Name	Resolution
Level-1	• 1A-GMI: GMI packet data transmitted by the satellite • 1B-GMI: GMI brightness temperatures • 1C-GMI: Calibrated GMI brightness temperatures • 1C-R: Common calibrated brightness temperatures collocated • 1C-constellation: Calibrated brightness temperatures for each passive-microwave instrument in the GPM constellation	• 1A-GMI: 8 km x 15 km (varies based on scan position), 16 orbits per day • 1B-GMI: Varies by Channel—16 orbits per day • 1C-GMI: Varies by Channel—16 orbits per day • 1C-R: Varies by Channel—16 orbits per day • 1C-constellation: Varies by satellite
Level-2	• 2A-GPROF-GMI: GMI single-orbit rainfall estimates • 2A-GPROF-constellation: Single-orbit rainfall estimates from each passive-microwave instrument in the GPM constellation • 2A-DPR: DPR Ka&Ku single orbit rainfall estimates • 2A-Ka: DPR Ka-only single orbit rainfall estimates • 2A-Ku: DPR Ku-only single orbit rainfall estimates • 2B-CMB: Combined GMI + DPR single orbit rainfall estimates	• 2A-GPROF-GMI: 8 km x 15 km (varies based on scan position), 16 orbits per day • 2A-GPROF-constellation: Varies by satellite • 2A-DPR: 5.2 km x 125 m—16 orbits per day • 2A-Ka: 5.2 km x 125 m—16 orbits per day • 2A-Ku: 5.2 km x 125 m—16 orbits per day • 2B-CMB: 5 km—16 orbits per day
Level-3	• 3-GPROF: GMI rainfall averages • 3-GPROF Constellation: Gridded rainfall estimates from each microwave imager in the GPM constellation • 3-DPR: DPR rainfall averages • 3-CMB: Combined GMI + DPR rainfall averages • IMERG: Rainfall estimates combining data from all passive-microwave instruments in the GPM Constellation (Early, Late, and Final)	• 3-GPROF: 0.25°, daily and monthly • 3-GPROF Constellation: 0.25°, daily and monthly • 3-DPR: 0.25°, daily and monthly • 3-CMB: 0.25° and 5°, daily and monthly • IMERG: 0.1°, 30 minute, daily, and monthly

Table 6.
A list of GPM datasets at GES DISC. More information is available in each dataset landing page.

GPM data products at GES DISC [25] are categorized also in three processing levels (**Table 6**). Like the TRMM era, a new multi-satellite, multi-retrieval product suite (IMERG) has been developed, with significant improvements in both spatial (0.1 x 0.1 deg.) and temporal (half hourly) resolutions over TMPA. There are three dataset categories in IMERG, Early, Late, and Final. The IMERG-Early provides near-real-time (latency: ~4 hours) global precipitation estimates, which are suitable for various research and applications such as flood watching. As more data are available, IMERG-Late (latency: ~12 hours) provides better estimates on precipitation than the Early. The IMERG-Final (latency: ~3.5 months) is a research-grade dataset that is bias corrected with ground gauge data from the Global Precipitation Climatology Centre (GPCC). **Figure 6** is an example of IMERG-Final, showing the accumulated rainfall of Hurricane Harvey during August 24–31, 2017.

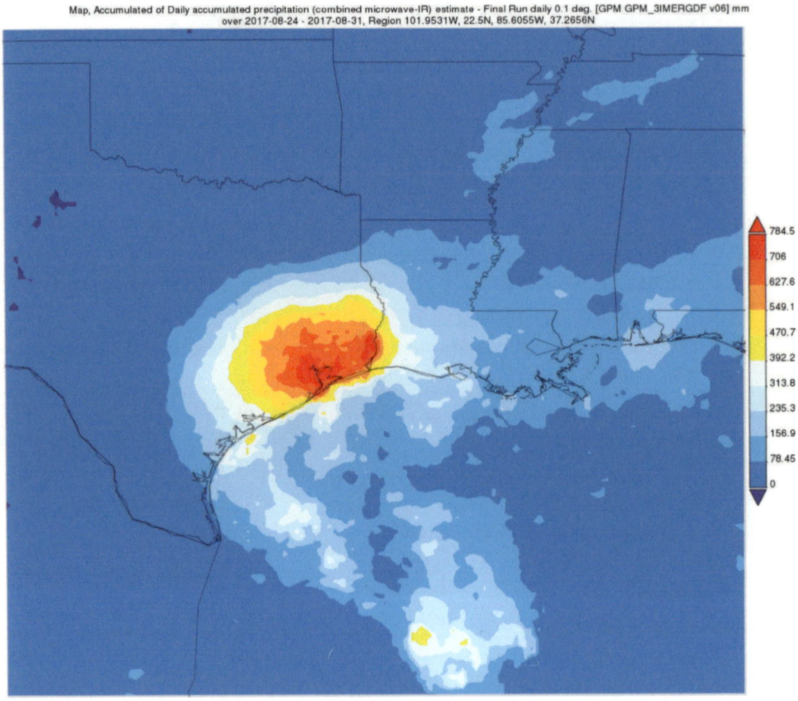

Figure 6.
Accumulated rainfall during August 24–31, 2017 from Hurricane Harvey. The map was generated with the GPM IMERG—Final daily dataset and Giovanni.

3.3 MERRA-2 dataset collection

Datasets from the Modern-Era Retrospective analysis for Research and Applications, Version 2 (MERRA-2) are developed by the NASA Global Modeling and Assimilation Office (GMAO) to place NASA's Earth Observing System (EOS) satellite observations in a climate context and to improve the representation of the atmospheric branch in the hydrological cycle from previous reanalysis or MERRA [26]. MERRA-2, available from 1980 onward, also includes the first long-term global aerosol reanalysis through assimilating satellite-based observations and representing their interactions with other physical processes in the Earth's climate system [26]. There are 95 product groups, and the file format is in NetCDF-4. Key meteorological variables in MERRA-2 for tropical cyclone studies such as wind, air temperature, and geopotential height are available at GES DISC. The spatial resolution is about 0.5 deg. x 0.625 deg. in the latitudinal and longitudinal directions, respectively, with 42 pressure levels and 72 model levels. MERRA-2 temporal resolutions range from hourly, 3 hourly, daily to monthly. Initial evaluation of MERRA-2 has been done and is available on the GMAO Web site [27].

4. Tools and data services at GES DISC

4.1 Tools

In research, data evaluation is often the first step to examine and understand a new physical dataset. Due to the complexity of satellite-based datasets, it is not an easy task to conduct such a task, especially for those without some prior knowledge about the dataset. Over the years, many tools have been developed by different organizations to facilitate such tasks. In this chapter, we describe two popular tools, GES DISC Giovanni and NASA GISS Panoply.

4.1.1 GES DISC Giovanni

Giovanni [4–7, 28] provides a simple and easy way to analyze and visualize more than 2000 satellite- and model-based physical parameters archived at GES DISC without downloading data and software. Variables of many well-known satellite missions and projects mentioned earlier such as TRMM, GPM, MERRA2, etc. are included in Giovanni.

Giovanni was first developed at the beginning of the TRMM era when TRMM TMPA datasets were available to the public [6]. Precipitation is a very popular variable and is used in many disciplines such as hydrology and agriculture. At that time, standard archived TMPA files were written in HDF4, a format that was not well known outside the remote sensing community. As a result, many TRMM users had difficulties to handle such format, which was a major barrier for TMPA data access and utilization. Recognizing this problem, scientists and software engineers at GES DISC worked closely with the TMPA product provider and developed a Web-based tool, the TRMM Online Visualization and Analysis System (TOVAS) [6]. With a Web browser, one can obtain average and accumulated rainfall maps as well as time series plots and Hovmöller diagrams in their areas of interest. Users can download results in several commonly used formats such as ASCII, which can be directly imported into Microsoft Excel for further processing. Later, several MODIS atmospheric products (e.g., aerosols, atmospheric water vapor) were added to TOVAS.

To meet an increasing demand for adding more analytical functions and variables, GES DISC developed Giovanni, allowing functions and variables to be added through a Web-based interface [4, 5]. In current version of Giovanni [7], more advanced information technologies have been implemented in the development, such as having all variables accessible in one Web interface, facet searching, sorting, provenance, etc. As of this writing, over 2000 variables from different Earth science disciplines are available and searchable in Giovanni, including datasets curated through other DAACs. More than 1300 referral research papers have been published by users around the world, with help from Giovanni. In short, Giovanni provides an easy way to evaluate and explore Earth science data at GES DISC.

With over 2000 variables in Giovanni, it is necessary to provide flexible search capabilities. Frist, users can type in key words such as IMERG and see variables only related to the key words. Often results from a search can contain many variables. For example, a search for precipitation returns 143 variables. To locate those of interest, one can sort the results based on source, spatial and temporal resolutions, begin or end dates. Facets have been developed to help narrow down search results, including disciplines, measurements, platform/instrument, spatial resolutions, temporal resolutions, wavelengths, etc.

Giovanni provides an interface for selecting date range. Users can either pick a date from the Web interface or type in their own date information. Likewise, users can draw their region of interest or type in the longitude and latitude coordinates in the interface, or they can select predefined shape files from a list including countries, land/sea masks, U.S. states, and major hydrological or watershed basins around the world.

There are 22 built-in analytical functions that are grouped into 5 categories based on their analysis types such as maps, time series, and comparisons. Once all the selections are done in the Web interface, a click on the "Plot Data" button will direct the user to the visualization result page where the user can find different options to fine tune their maps or plots. The output page provides a browse history in which users can return back to the input Web interface, download the results in graphic or NetCDF formats, or visit the lineage to see the provenance information or download data in each process. **Figure 6** is an example of using Giovanni to generate the accumulated rainfall from Hurricane Harvey during August 24–31, 2017.

Giovanni training materials have been developed over the years. The Giovanni user guide [29] is available through the help button along with release notes, browser compatibility, and known issues. Users can also visit YouTube for Giovanni-related How-to videos [30]. The NASA Applied Remote Sensing Training (ARSET) project also provides materials for Giovanni online training [31], used in live webinars that are free of charge for users around the world. If users have questions or suggestions about Giovanni, they can submit them thorough the feedback button in the landing page and a staff member at GES DISC will provide assistance. Acknowledgment policy is also available at the bottom of the output page.

4.1.2 NASA GISS Panoply

Although users can access over 2000 variables through Giovanni, all these variables are in Level-3. While adding data variables from other levels in Giovanni are being considered, users can use Panoply [32], developed by NASA Goddard GISS, to view Level-2 and Level-3 data. Panoply is another powerful tool for viewing NASA data. Panoply can be installed in several platforms and operating systems (e.g., macOS, Windows), requiring Java 8 or later version installed in their systems. Most datasets archived at GES DISC can be viewed by Panoply. Once Panoply is installed, there are several ways to import data to Panoply. If NetCDF-3 or NetCDF-4 is the default setting for opening files in the system, Panoply will automatically open the file when you download it from GES DISC. Popular Level-2 and Level-3 datasets are often available in OPeNDAP and GrADS Data Server (GDS). Users can directly use their dataset links in Panoply to visualize the variables. Although analytical functions in Panoply are quite limited, it is so far an easy way to visualize datasets that are currently not available in Giovanni (**Figures 3, 4** and **5a**).

4.2 Data services

Data services are essential for data archive centers. The totally redesigned GES DISC Web site [3] (**Figure 1**) makes datasets, documents, and help information easy to find. Due to a large volume of datasets and information archived at GES DISC, a search capability is necessary to facilitate dataset discovery and exploration. From the search box (**Figure 1**), users can search datasets and information in the following categories: data collections, data documentation, alerts, FAQs, glossary, How-to's, image gallery, news, and tools. The category for data collections is the default since many users come to GES DISC for datasets. Users can also browse data by category, including subject, measurement, source, processing level, project, temporal resolution, and spatial resolution. As of this writing, the GES DISC archives more than ~2.3 PB data with over 117 million files. Over 2.4 billion files have been distributed with data volume over 23 PB.

After a user types in a key word in the search box (**Figure 1**), the search results are listed. Faceting and sorting are available for identifying datasets of interest, similar to those in Giovanni. For example, a search for IMERG returns 15 datasets and users can use Version to sort different versions and find out the datasets of the latest version. "Get Data" or "Subset/Get Data" right below a dataset name provides a direct link to the data download interface. A click on a dataset name leads to the dataset landing page with more information on dataset summary, data citation, documentation, and more data access methods including links to online archive,

Giovanni, Web services such as OPeNDAP, GDS, and THREDDS (the Thematic Real-time Environmental Distributed Data Services) available for some popular datasets.

Most NASA datasets are global coverage. Many study areas are either local or regional; therefore, spatial subsetting is important to reduce download volumes, permitting the user more time to do research. Spatial subsetting is available for a large number of datasets at GES DISC. Users can use a computer mouse to drag an area of interest or type in the geolocation coordinates in the subsetter interface. Some subsetters can do more than spatial subsetting such as parameter subsetting, i.e., selecting wanted variables from a list, which can also help reduce data to be downloaded. For MERRA-2 data, the subsetter can also regrid the original MERRA-2 data into different grid structures with a list of interpolation methods (e.g., bilinear, bicubic). Furthermore, the MERRA-2 subsetter can subset data at pressure levels. NetCDF-4 is available for many datasets at GES DISC, which can be handled by many software packages or tools such as Panoply, ArcGIS.

Level-2 data subsetting is available for popular and high-volume datasets such as the Level-2 GPM dual-frequency product from DPR (2ADPR) that provides general characteristics of precipitation, correction for attenuation, profiles of precipitation water content, rain rate, as well as particle size distributions of rain and snow. The dual-frequency observations from DPR provide better estimates of rainfall and snowfall rates than the single-frequency TRMM PR [33] with additional information for particle size and melting layer height from the Ka band. Variables in 2ADPR are available in all three scan modes of DPR: a) normal scans (NS), b) matched scans (MS) (**Figure 5a**), c) and high sensitivity (HS), and their swath sizes range from 120 km to 245 km. Each file contains over 400 variables with size close to 300 MB. The subsetting service, developed at GES DISC, provides both variable and spatial subsetting capabilities, which help reducing the file size by several orders of magnitude, depending on selections. Three spatial subsetting capabilities (**Figure 5b**) are currently available: a rectangular latitude/longitude box, a circle, and a point. Users can pick one of them in the interactive subsetting Web interface and create an area or point of interest. These spatial and parameter subsetting capabilities facilitate ground validation and evaluation activities. For example, users can pull DPR data over a time period for a location where a rain gauge is located for validation or evaluation.

For decades, the GES DISC has archived and distributed a large amount of Earth science data, information, and services to diverse communities including the tropical cyclone community. From searching, discovering to assessing such "Big Data," i.e., heterogeneous and immense scientific data (particularly, satellite or model products) in order to timely and properly examine and assess those natural devastating weather events with an imminent goal for better understanding their natures and reducing the resultant disaster risk, it has, nonetheless, become a daunting task for science researchers and application users (and decision makers).

Aiming to substantially assist our users in their online effectively (i.e., quickly and properly) acquiring the data they want and/or need a "one-stop shop" with a minimum effort from our large data collection for their investigating and assessing the targeted disastrous weather such as hurricanes, the GES DISC has recently developed a value-added and knowledge-based data service prototype by preparing/presenting the "List" of relevant data and the pertinent resources accordingly. Such a data service framework, termed as "Datalist" (currently containing "Hurricane Datalist" only), which basically consists of suites of annotated Web addresses (URLs) that point to the proper and relevant data and resources. **Figure 7** (concept based on [34, 35]) shows a basic workflow of how a user can online, via accessing Hurricane Datalist, acquire and assess the respective datasets (down to the variable level such as wind and air temperature, etc.) or services (such as Subsetting and Giovanni) they want or need relevant to their targeted hurricane event, e.g., Hurricane Sandy (October 22–29, 2012) over the US continent and coast area at one stop. Basically, through visiting the Hurricane Datalist page (**Figure 7**), users can readily choose and apply those handy "Subsetting" options of (1) refining data temporal range; (2) selecting spatial domain; (3) choosing targeted variables; and (4) acquiring and downloading the data they want and/or need. Moreover, a useful "window shopping" service is offered to users, allowing them to utilize Giovanni

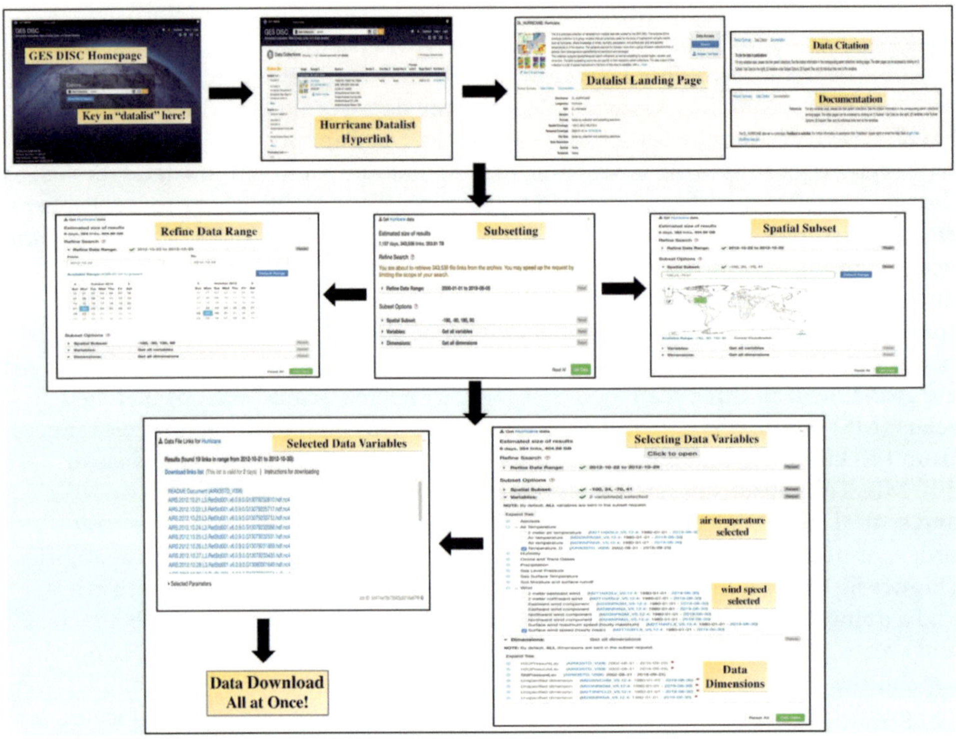

Figure 7.
Flow chart of discovering and accessing data sets and variables, e.g., hurricane Sandy (October 22–29, 2012) via hurricane Datalist.

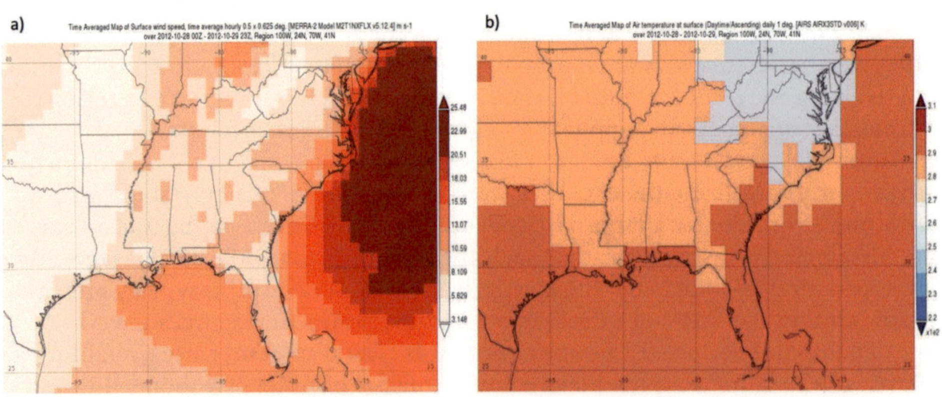

Figure 8.
(a) MERRA-2 wind speeds, (b) AIRS air temperature during October 28–29, 2012 involving hurricane Sandy (October 22–29, 2012).

to plot and view their interested variables pre-"data downloading" that help them make proper decisions. **Figure 8** shows two maps produced with Giovanni for the two sampled data variables, i.e., the MERRA-2 wind speeds (**Figure 8a**) and the AIRS air temperature (**Figure 8b**).

Users must register the NASA Earthdata login system before downloading data from any NASA data center (including GES DISC). A help feature [3] is available to assist users when they have questions about data and services, which can be very helpful for those who are not familiar with NASA satellite datasets and may not know where to begin. Using this feature, GES DISC staff and supporting scientists can guide users regarding questions related to datasets, tools, documentation, and services. FAQs and How-To recipes are also available and searchable.

5. Case studies

5.1 Case 1: Evaluation of MERRA-2 precipitation during hurricane Katrina landfall

Evaluation (e.g., comparison) of a dataset prior to download is important to understand (for example) any biases or systematic differences in datasets, which is quite common for remote sensing and model datasets. Over oceans, few in situ observations are available, especially for precipitation, making it very difficult to assess their biases. MERRA-2 datasets provide over 39 years of continuous analysis ranging from hourly to monthly, as mentioned earlier, and can be used to study events and environmental changes (e.g., trends) in tropical oceans and other regions as well.

There are two types of precipitation parameters in MERRA-2: a) precipitation from the atmospheric model (variable PRECTOT in the MERRA-2 dataset collection) and b) observation-corrected precipitation (variable PRECTOTCORR) [36, 37]. Observational data are introduced in the latter parameter due to considerable errors that propagate into land surface hydrological fields and beyond [38].

Bosilovich et al. [37] have conducted a general evaluation of MERRA precipitation estimates, including precipitation climatology, interannual variability, diurnal cycle, Madden-Julian Oscillation (MJO) events, global water cycle, and U.S. summertime variability. Although the preliminary evaluation provides a basic understanding of the MERRA-2 precipitation products, evaluation for extreme weather events is still needed to better understand MERRA-2 precipitation behavior and characteristics.

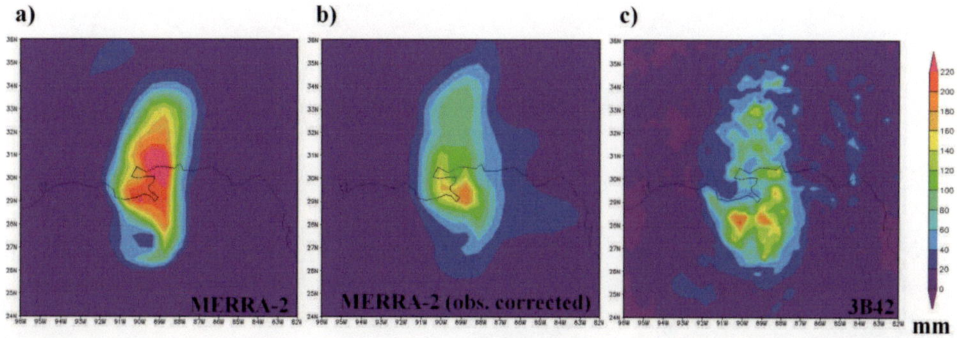

Figure 9.
Daily precipitation total (in mm) during hurricane Katrina landfall on August 29, 2005 from (a) MERRA-2 modeled precipitation; (b) observation-corrected precipitation; and (c) TMPA 3B42.

Figure 9 shows 3 daily precipitation maps from the MERRA-2 modeled, observation-corrected, and bias-corrected TMPA 3B42 precipitation products on Aug. 29, 2005, when the deadly Hurricane Katrina, as currently ranked as the 3rd most intense landfalling hurricane in the U.S. history, made a landfall near New Orleans, Louisiana. Katrina claimed at least 1245 lives, making it the deadliest U.S. hurricane since the Hurricane Okeechobee in 1928. Apart from the obvious difference in the dataset spatial resolutions (~0.5° in MERRA vs. 0.25° in 3B42), it is seen that large differences exist among three precipitation products (**Figure 9**). **Figure 9a** shows that the modeled precipitation has the largest systematic differences against TMPA 3B42 (**Figure 9c**) in terms of intensity and structure. Significant differences still exist even

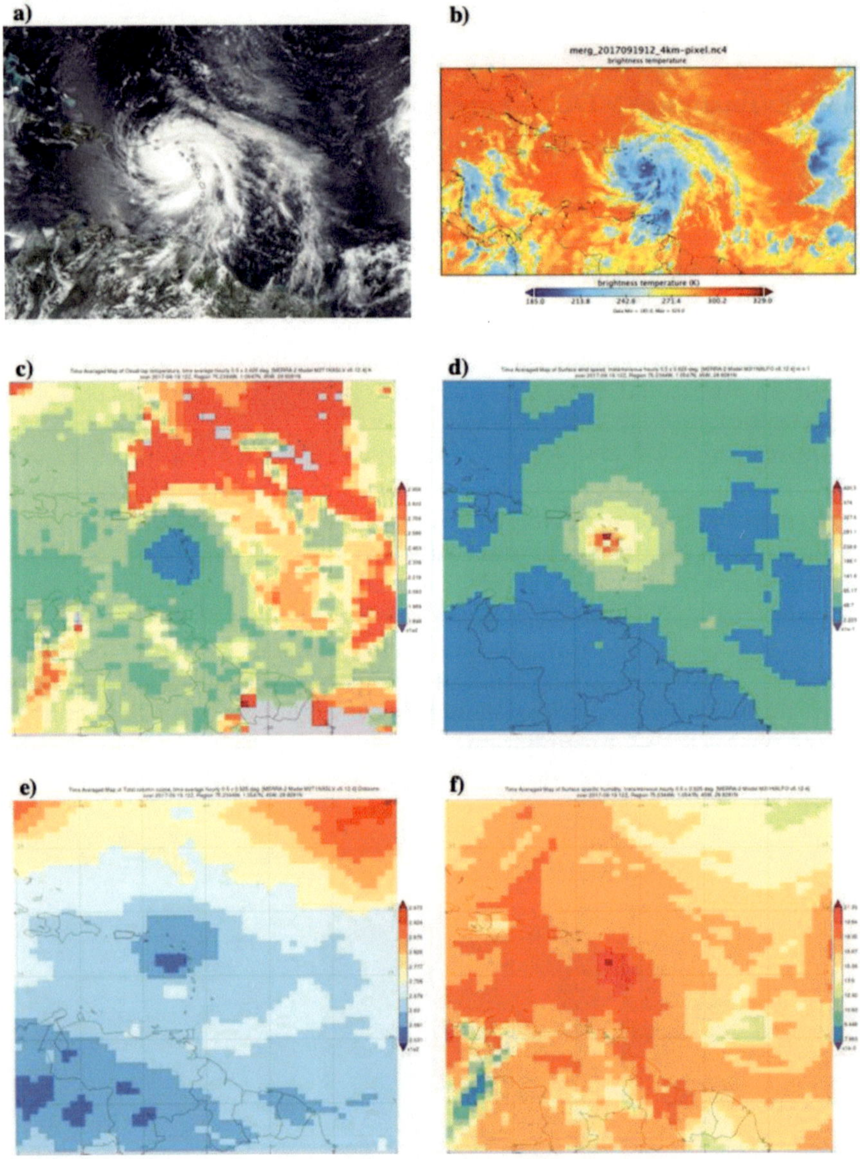

Figure 10.
Sample images of hurricane Maria at 12Z September 19, 2017 from different datasets and services: (a) true color image from Suomi NPP in NASA Worldview; (b) NOAA/CPC merged IR from the GES DISC archive; (c) MERRA-2 cloud top temperature; (d) MERRA-2 surface wind speed; (e) MERRA-2 total column ozone; and (f) MERRA-2 surface specific humidity.

after observational data were applied for bias correction (**Figure 9b** and **c**). More importantly, the centers of heavy rainfall are mismatched in all datasets. In TMPA, there are two centers and all are over the ocean (**Figure 9c**). By contrast, the center from the modeled is located on land (**Figure 9a**). The center from the observation-corrected is between the modeled and TMPA (**Figure 9b**). Also, it seems that both MERRA-2 precipitation products do not do well in the light rain regions in **Figure 9**. The overestimation issue in this case seems to be consistent with the evaluation conducted by Bosilovich et al. [37]; however, other important issues (e.g., the heavy rain centers in wrong places) do exist.

5.2 Case 2: Acquiring Earth science data measurements all at once to study hurricane events

Collecting and evaluating data are important activities for tropical cyclone research. A one-stop shop for acquiring these activities can save time and is very desirable. However, such system is still being developed and there are many obstacles to overcome. This case study is to demonstrate, with existing tools, one can conduct such activities.

With Giovanni, users can explore over 2000 satellite- and model-based variables. For example, Hurricane Maria is a deadly Category-5 hurricane and caused a heavy damage on local economy in island countries in the Caribbean. Before using Giovanni, one can obtain the true color satellite image from NASA Worldview (**Figure 10a**) and the IR image from the merged IR dataset at the GES DISC and Panoply (**Figure 10b**). The latter provides uninterrupted IR data at 4-km spatial resolution available every 30 minutes, which is very helpful for tracking the hurricane evolution. There are many variables in Giovanni from TRMM, GPM, MERRA-2, etc. Once the beginning and ending times as well as the geolocation are decided, one can input such information in Giovanni. The next step is to select variables of interest. In this case, four MERRA-2 variables are selected: (a) cloud top temperature (**Figure 10c**); (b) surface wind speed (**Figure 10d**); (c) total column ozone (**Figure 10e**); and (d) surface specific humidity (**Figure 10f**). In **Figure 10b** and **c**, large differences in cloud top temperatures exist between the merged IR dataset and MERRA-2. For example, the MERRA-2 cloud top temperatures appear to be cooler than those of the merged IR north of the coast of Venezuela. High wind speeds are found near the hurricane center (**Figure 10d**) where low total column ozone (**Figure 10e**) and high surface specific humidity (**Figure 10f**) are located.

6. Summary and future plans

In this chapter, we present an overview of basic datasets and services at GES DISC for tropical cyclone research. The collection at GES DISC includes datasets from major NASA satellite missions (e.g., TRMM, GPM) and projects (e.g., MERRA-2, GPCP) with emphasis on precipitation, hydrology, atmospheric composition, atmospheric dynamics, etc. The GES DISC provides user-friendly data services to facilitate data evaluation and download, including a) Giovanni, an online visualization and analysis tool for access over 2000 variables without downloading data and software; b) data subsetting services that allow spatial and variable subsetting of Level-2 and Level-3 datasets; c) different data access methods (online archive, OPeNDAP, GDS, THREDDS) and data formats for a wide variety of users with various technical expertise in handling complex remote sensing datasets; d) information about documentation and data citation; and e) user services to answer data or service-related questions.

We present two case studies to show how our datasets can be used in tropical cyclone research. In the first case, three different precipitation datasets were compared during the landfall of Hurricane Katrina. Results show that large differences exist among the three datasets. The MERRA-2 modeled precipitation has a large wet bias compared to the other two datasets. The areas of heavy precipitation for both modeled and observation-corrected are different from those of 3B42. The results suggest that these differences need to be considered when using the model precipitation products. The second case is an example of exploring different satellite- and model-based variables from existing data services and tools such as Worldview, Panoply, and Giovanni.

Two main areas are focused in future plans: datasets and services. First, as mentioned earlier, NASA Earth data are archived at 12 discipline-oriented data centers across the United States. Datasets at other NASA data centers are also important for tropical cyclone research. For example, the Physical Oceanography DAAC at JPL archives key measurements for tropical cyclone research such as satellite-based ocean surface wind, sea surface temperature, etc. NASA airborne and field campaign datasets for hurricane research, archived at the Global Hydrology Resource Center, play an important role in product validation and case studies. One challenge is to facilitate data discovery and access across the DAACs because relevant datasets are located in different or multiple centers and each center has its unique Web interface for ordering data as well as tools for customized analysis and visualization, which may create problems to some users. It would be more user friendly and efficient to have a one-stop Web interface with data services for acquiring datasets from different data centers. Prototypes have been developed specifically for tropical cyclone research, but they have not been fully integrated into operation and only limited datasets are available; therefore, the datasets can be incomplete. Currently, only NASA Earthdata allows searching datasets from the 12 NASA data centers with very limited data services available. As well, tools can be consolidated and further developed to facilitate access to datasets at different data centers. More on this is to be elaborated in the services.

Analysis-ready data can save time and expedite research and discovery because data are pre-processed at a data center based on research needs. These data are friendly to user's written analysis software or publicly available software packages or tools. For example, time series data subsetting can be a challenging issue for some communities (e.g., hydrology). NASA data are file based, one-time step per file with complex data structures containing few or more variables, which is not optimal for time series data access [39]. To make data analysis ready, data may need to be re-organized for efficient access, such as is demonstrated by the data rods concept [39]. Increasingly, machine learning (ML) and artificial intelligence (AI) algorithms are being used in many areas including tropical cyclone research. Training data play an important role in both ML and AI development and applications. Making training data from collections at DAACs analysis ready (e.g., providing event-based data subsets) can save time for downloading data and processing.

The NASA's Earth Observing System Data and Information System (EOSDIS) cloud evolution [40] is a project to deploy NASA Earth science data and services into a commercial cloud environment to improve data accessibility and serviceability across all NASA DAACs. Prototypes are currently in development to demonstrate cross-DAAC data discovery, access, and servicing. Some datasets are moving natively and operationally into cloud environments as of this writing, such as those from the upcoming Surface Water and Ocean Topography (SWOT). Selected GES DISC datasets are scheduled to be deployed into a cloud environment in 2020, including the IMERG and MERRA-2 collections. Placing the EOSDIS archive collectively in the cloud will, for the first time, place NASA Earth Observation (EO) data "close to compute" and improve management and accessibility of these data while also expediting science discovery for data users. This will also enable large-scale data analytics for data users, especially allow more efficient use data from multiple DAACs.

As mentioned earlier, NASA Earth science data collected from satellites, model assimilation, airborne missions, and field campaigns are large, complex, and evolving. It can be a daunting task to obtain data from different centers with different interfaces or tools. Data services increasingly play an important role to facilitate data discovery, access, and exploration. Data services at GES DISC are evolving as well. For example, our currently predefined Datalist by no means can contain variables in all possible research topics in hurricane research. User-defined Datalist would allow users to define their own Datalist is the next thing to be developed. The user-defined Datalist not only can save time for dataset search but also can be sharable to other collaborators or scientists.

Since tropical cyclone research can be categorized based on different spatial and temporal scales. For mesoscale or synoptic scale, event case studies are heavily conducted in tropical cyclone research, requiring that an event information can be easily retrieved from a database (e.g., Hurricane best track data or HURDAT) in order to locate relevant datasets for subsetting, analysis, and visualization with that information (spatial and temporal constrains). Furthermore, datasets can also be searched and located based on other information such as track, intensity, or criteria set by users. At present, neither Giovanni nor the GES DISC Web interface has such capabilities. Adding such event databases can also create data subsets for other research activities such as wild fires, volcanic eruptions, heat waves, snow storms, and nor'easters.

Integration, analysis, and visualization for NASA Level-2 and airborne products are challenging because there are a lot of data-related issues such as formats, structures, terminology, etc. Furthermore, integration of Level-2 and Level-3 products is also needed since model and multi-satellite products (MERRA-2, IMERG) are gridded Level-3 products. The first challenge is to be able to locate (then collocate) available datasets (e.g., swath) for an event and subset the data for the area of interest. The latter work has been done for some GPM products at GES DISC, as mentioned earlier. Customized development is necessary to deal with different data structures from different satellite missions as well as airborne and field campaigns, requiring a close collaboration among data centers. At present, there is no Level-2 dataset in Giovanni. Adding Level-2 datasets and airborne data in Giovanni is important to significantly expand the capabilities of Giovanni in tropical cyclone research because observations are very limited over vast tropical oceans, and all these available observations are important for a wide variety of research activities, regardless. Participation and feedback from users or stakeholders always play a key role to ensure development results to be user friendly and useful.

For climate scale, data services need to provide essential information and datasets including climatological datasets (e.g., sea surface temperature, ocean surface wind speed), anomalies, trends, etc. to help researchers to understand changes and trends in environmental conditions over tropical oceans where tropical cyclones are born and developing. Long-term datasets are important, such as MERRA-2 datasets provide over 39 years of global assimilation analysis (1980–present), which is suitable for generating climatological datasets. Giovanni provides on-the-fly generation of climatology and time series plots for several key datasets, such as MERRA-2, TMPA, and IMERG, for tropical cyclone research.

Acknowledgements

We thank scientists and engineers at GES DISC for their contributions to data management, distribution, and development of data services. We also thank scientific investigators and many users for their feedback and suggestions that improve our data services. GES DISC is funded by NASA's Science Mission Directorate.

Author details

Zhong Liu[1,2]*, David Meyer[1], Chung-Lin Shie[1,3] and Angela Li[1]

1 NASA Goddard Earth Sciences Data and Information Services Center (GES DISC), USA

2 George Mason University, USA

3 University of Maryland Baltimore County, USA

*Address all correspondence to: zhong.liu@nasa.gov

References

[1] Kidder SQ, von der Haar TH. Satellite Meteorology - an Introduction. San Diego, CA, USA: Academic Press; 1995, 1995. p. 466. ISBN 012 406430 2

[2] Datta A. 2019, A Brief History of Weather Satellites. Available from: https://www.geospatialworld.net/blogs/a-brief-history-of-weather-satellites/ [Accessed: 11 August 2019]

[3] NASA GES DISC. The NASA GES DISC. 2019. Available from: https://disc.gsfc.nasa.gov/ [Accessed: 11 August 2019]

[4] Acker JG, Leptoukh G. Online analysis enhances use of NASA earth science data. Eos Transaction AGU. 2007;**88**(2):14-17. DOI: 10.1029/2007EO020003

[5] Berrick SW et al. Giovanni: A web service workflow-based data visualization and analysis system. IEEE Transactions on Geoscience and Remote Sensing. 2009;**47**(1):106-113. DOI: 10.1109/TGRS.2008.2003183

[6] Liu Z et al. Online visualization and analysis: A new avenue to use satellite data for weather, climate, and interdisciplinary research and applications. In: Levizzani V et al., editors. Measuring Precipitation from Space: EURAINSAT and the Future, Adv. Global Change Res. Ser. Vol. 28. New York: Springer; 2007. pp. 549-558. DOI: 10.1007/978-1-4020-5835-6

[7] Liu Z, Acker J. Giovanni: The Bridge Between Data and Science, Eos, 98. 2017, https://doi.org/10.1029/2017EO079299. Published on 24 August 2017

[8] Nature. Scientific Data. 2019. Available from: https://www.nature.com/sdata/ [Accessed: 11 August 2019]

[9] NASA, New Data from Old Satellites: A Nimbus Success Story. 2019, Available from: https://earthdata.nasa.gov/new-data-from-old-satellites-a-nimbus-success-story [Accessed: 11 August 2019]

[10] NASA NSIDC, The Nimbus Data Rescue Project. 2019, Available from: https://nsidc.org/data/nimbus [Accessed: 11 August 2019]

[11] Janowiak JE, Joyce RJ, Yarosh Y. A real-time global half-hourly pixel-resolution infrared dataset and its applications. Bulletin of the American Meteorological Society. 2001;**82**(3):205-217

[12] NASA GES DISC, NOAA/NCEP/CPC Half Hourly 4km Global (60S - 60N) Merged IR. 2019. Available from: https://disc.gsfc.nasa.gov/datasets/GPM_MERGIR_V1/summary [Accessed: 11 August 2019]

[13] Huffman GJ, Bolvin DT, Braithwaite D, Hsu K, Joyce R, Kidd C, et al., IMERG Algorithm Theoretical Basis Document (ATBD) Version 06. 2019. Available from: https://pmm.nasa.gov/sites/default/files/document_files/IMERG_ATBD_V06.pdf [Accessed: 11 August 2019]

[14] Joyce RJ, Janowiak JE, Arkin PA, Xie P. CMORPH: A method that produces global precipitation estimates from passive microwave and infrared data at high spatial and temporal resolution. Journal of Hydrometeorology. 2004;**5**:487-503

[15] Hou AY et al. The global precipitation measurement mission. Bulletin of the American Meteorological Society. 2014;**95**:701-722. DOI: 10.1175/BAMS-D-13-00164.1

[16] NASA GSFC PPS and X-Cal Working Group. Algorithm Theoretical Basis Document (ATBD) NASA Global Precipitation Measurement Level 1C

Algorithms Version 1.8. 2017. Available from: https://pps.gsfc.nasa.gov/Documents/L1C_ATBD.pdf [Accessed: 11 August 2019]

[17] Garstang M, Kummerow CD. The Joanne Simpson special issue on the tropical rainfall measuring mission (TRMM). Journal of Applied Meteorology. 2000;**39**:1961

[18] Liu Z, Ostrenga D, Teng W, Kempler S. Tropical rainfall measuring Mission (TRMM) precipitation data Services for Research and Applications. Bulletin of the American Meteorological Society. 2012;**93**(9):1317-1325. DOI: 10.1175/BAMS-D-11-00152.1

[19] NASA. NASA Data Processing Levels. 2019. Available online: https://earthdata.nasa.gov/collaborate/open-data-services-and-software/data-information-policy/data-levels [Accessed: 11 August 2019]

[20] Huffman GJ et al. The TRMM multisatellite precipitation analysis (TMPA): Quasi-global, multiyear, combined-sensor precipitation estimates at fine scales. Journal of Hydrometeorology. 2007;**8**:38-55

[21] Huffman GJ, Adler RF, Bolvin DT, Nelkin EJ. Chapter 1: The TRMM multi-satellite precipitation analysis (TAMPA). In: Hossain F, Gebremichael M, editors. Satellite Rainfall Applications for Surface Hydrology. Dordrecht, Netherlands: Springer Verlag; 2010, ISBN: 978-90-481-2914-0, 3-22

[22] Huffman GJ, Bolvin DT. TRMM and Other Data Precipitation Data Set Documentation. 2013. Available from: ftp://meso-a.gsfc.nasa.gov/pub/trmmdocs/3B42_3B43_doc.pdf [Accessed: 11 August 2019]

[23] NASA GHRC. Global Hydrology Resource Center. 2019, Available from: https://ghrc.nsstc.nasa.gov/home/ [Accessed: 11 August 2019]

[24] Soula S. Lightning and precipitation. In: Betz HD, Schumann U, Laroche P, editors. Lightning: Principles, Instruments and Applications. Dordrecht. ISBN: 978-1-4020-9078-3: Springer; 2009. DOI: 10.1007/978-1-4020-9079-0_20

[25] Liu Z, Ostrenga D, Vollmer B, et al. Global precipitation measurement mission products and services at the NASA GES DISC. Bulletin of the American Meteorological Society. 2017;**98**(3):437-444. DOI: 10.1175/bams-d-16-0023.1

[26] Gelaro R et al. The modern-era retrospective analysis for research and applications, version 2 (MERRA-2). Journal of Climate. 2017;**30**:5419-5454. DOI: 10.1175/JCLI-D-16-0758.1

[27] NASA GMAO. Modern-Era Retrospective analysis for Research and Applications, Version 2. 2019. Available from: https://gmao.gsfc.nasa.gov/reanalysis/MERRA-2/ [Accessed: 11 August 2019]

[28] NASA GES DISC. Giovanni. 2019. Available from: https://giovanni.gsfc.nasa.gov/giovanni/ [Accessed: 11 August 2019]

[29] NASA GES DISC. Giovanni User Guide. 2019. Available from: https://giovanni.gsfc.nasa.gov/giovanni/doc/UsersManualworkingdocument.docx.html [Accessed: 11 August 2019]

[30] NASA GES DISC. Giovanni Tutorials on YouTube. 2019. Available from: https://www.youtube.com/results?search_query=nasa+giovanni+tutorial [Accessed: 11 August 2019]

[31] NASA ARSET. Applied Remote Sensing Training. 2019. Available from: https://arset.gsfc.nasa.gov/ [Accessed: 11 August 2019]

[32] NASA GISS. Panoply. 2019. Available from: https://www.giss.nasa.gov/tools/panoply/download/ [Accessed: 11 August 2019]

[33] Iguchi T, Seto S, Meneghini R, Yoshida N, Awaka J, Le M, et al. GPM/DPR Level-2 Algorithm Theoretical Basis Document (ATBD). 2017. Available from: https://pmm.nasa.gov/sites/default/files/document_files/ATBD_DPR_201708_whole_1.pdf [Accessed: 11 August 2019]

[34] Shie C-L, Teng WL, Liu Z, Hearty T, Shen S, Li A, et al. Data List - Specifying and Acquiring Earth Science Data Measurements All at Once, the 2016 AGU Fall Meeting, 12-16 December 2016, San Francisco, CA; 2016

[35] Li A, Hegde M, Bryant K, Seiler E, Shie C-L, Teng WL, et al. Data List: A Value-Added Service to Enable Easy Data Selection, the 2016 AGU Fall Meeting, 12-16 December 2016, San Francisco, CA; 2016

[36] Reichle RH, Liu Q. Observation-Corrected Precipitation Estimates in GEOS-5. NASA/TM–2014-104606, Vol. 35; 2014

[37] Bosilovich M, Akella S, Coy L, Cullather R, Draper C, Gelaro R, et al. Techinical Report Series on Global Modeling and Data Assimilation. In: Koster R, editor. Greenbelt, MD, USA: NASA Goddard Space Flight Center; 2015. Vol. 43. Available from: http://gmao.gsfc.nasa.gov/pubs/tm/docs/Bosilovich803.pdf [Accessed: 11 August 2019]

[38] Rienecker MM, Suarez MJ, Gelaro R, Todling R, Bacmeister J, Liu E, et al. MERRA: NASA's modern-era retrospective analysis for research and applications. Journal of Climate. 2011;**24**:3624-3648. DOI: 10.1175/JCLI-D-11-00015.1

[39] Teng W, Rui H, Strub R, Vollmer B. Optimal reorganization of NASA earth science data for enhanced accessibility and usability for the hydrology community. Journal of the American Water Resources Association (JAWRA). 2016;**52**(4):825-835. DOI: 10.1111/1752-1688.12405

[40] NASA. NASA Cloud Evolution. 2019. Available from: https://earthdata.nasa.gov/eosdis/cloud-evolution [Accessed: 11 August 2019]

Chapter 7

Climate Models Accumulated Cyclone Energy Analysis

Sullyandro Oliveira Guimarães

Abstract

Looking at the connection between tropical cyclones and climate changes due to anthropogenic and natural effects, this work aims for information on understanding and how physical aspects of tropical cyclones may change, with a focus on accumulated cyclone energy (ACE), in a global warming scenario. In the present climate evaluation, reasonable results were obtained for the ACE index; the Coupled Model Intercomparison Project Phase 6 (CMIP6) models with lower horizontal and vertical resolution showed more difficulties in representing the index, while Max Planck Institute model demonstrated ability to simulate the climate with more accurate, presenting values of both ACE and maximum temperature close to NCEP Reanalysis 2. The MPI-ESM1-2-HR projections suggest that the seasons and their interannual variations in cyclonic activity will be affected by the forcing on the climate system, in this case, under the scenario of high GHG emissions and high challenges to mitigation SSP585. The results indicate to a future with more chances of facing more tropical cyclone activity, plus the mean increase of 3.1°C in maximum daily temperatures, and more heavy cyclones and stronger storms with more frequency over the North Atlantic Ocean may be experimented, as indicated by other studies.

Keywords: climate change, accumulated cyclone energy, SSP585 scenario, tropical cyclones, CMIP6

1. Introduction

The challenge of connecting climate change to tropical cyclones (TCs) lies in determining that a change has occurred given natural variability whether by significant changes in climate forcing such as greenhouse gases (GHGs) or aerosols or by the sum of both natural and anthropogenic factors.

Tropical cyclone activity has complex characteristics that make it difficult to achieve robust future projections. The onset, duration, intensity, and phenomenology associated with these storms carry many uncertainties in numerical modeling, due to limitations of models to represent local/micro-scale physical processes and tangents to the computational aspect in simulating the climate of long periods, from decades to centuries.

Changes in natural variability, volcanic emissions, and solar activity have made a small contribution to the changes in climate over the last century [1, 2]. The natural cycles observed in climate records do not explain the increases in the heat content of the atmosphere, ocean, or cryosphere since the industrial age [3–6].

IntechOpen

Earth's climate has been affected by changes in factors that control the amount of energy entering and leaving the atmosphere. These factors, known as radiative forcings, include changes in albedo through land use and cover, greenhouse gases, and aerosols. The increase in the concentration of greenhouse gases by emissions from human activities is the largest of these radiative forcings. By absorbing longwave radiation emitted by Earth and redirecting it equally in all directions, greenhouse gases increase the amount of heat retained in the climate system, warming the planet [2, 7–9].

A comparison of a model's simulation of internal multidecadal climate variability with the observed increase in an Atlantic hurricane rapid intensification metric (1982–2009) finds a highly unusual behavior in the metric result and is consistent with the long-term response sign expected by the model to anthropogenic forcing [10]. In the same direction, the 2018 US National Climate Change Assessment reports that decreases in air pollution and increases in GHGs have contributed to increases in Atlantic hurricane activity since 1970 [11].

There is growing evidence of a significant increase in the TC's proportion that become major hurricanes, although the frequency of TCs has remained roughly constant in recent decades [12–17]. A recent study showed that in the central and eastern tropical Atlantic basin during 1986–2015, the 95th percentile of 24 h intensity changes increased significantly [18]. The intensification rate of intensifying storms, another metric that is not dependent on TC frequency, exhibited significant growth during 1977–2013 in the West Pacific basin [19]. In both studies, the large-scale environment became more conducive to TC intensification over time. Areas with increases in potential intensities [20] and the largest increase in sea surface temperatures (SSTs) seem to be located with the largest positive changes in intensification rates.

How future anthropogenic warming can affect TC is an important issue, mainly due to the large social impacts they can cause [21], as discussed in previous reports of the Intergovernmental Panel on Climate Change (IPCC) [22] and World Meteorological Organization (WMO) [23].

The IPCC-AR5 [24] concludes for a 2°C global warming that there is more than 66% likelihood to the TC rainfall rates increase in the future and accompanying increase in atmospheric moisture content. Modeling studies on average indicate increase rainfall rates averaged within about 100 km of the storm by 10–15%. The TC intensities increase on average (1–10%), which would imply an even larger increase of percentage in the destructive potential per storm, assuming no reduction in storm size (responses to anthropogenic warming are uncertain).

The future projection for the global number of Category 4 and 5 storms is likely to increase due to anthropogenic warming over the twenty-first, but there is less confidence since most modeling studies project a decrease (or little change) in the overall frequency of all combined TC [24].

Links between climate and tropical cyclones were analyzed in [25], with a good understanding of the relationship at various time scales, with significant trends observed for cyclone intensity and frequency over the past decades over Atlantic. Most climate models simulate fewer tropical cyclones and stronger storms, with increase in precipitation rates. Further sea level rise is likely to increase storm threats, with studies of combined effects of floods and storms projecting that increases are due to global warming [26].

Given the importance of tropical cyclone study, and how changes induced by human actions in the terrestrial system may affect such phenomena, the aim of this study is to evaluate simulations of global numerical models of the Coupled Model Intercomparison Project Phase 6 (CMIP6) [27], by representing the recent past, and thus access future projections that may occur and indicate trends of changes in cyclone events.

2. Accumulated cyclone energy (ACE) approach

ACE uses the maximum wind speed over time to quantify hurricane activity by season, defined as the sum of the squares of the maximum wind speeds at 6-h intervals, considering the time while the hurricane is at tropical storm strength or greater [28]. As kinetic energy is proportional to the square of velocity, ACE is a value proportional to the energy of the system, by adding together the energy per some interval of time.

A review by [29] evaluates different hurricane indexes, indicating ACE as a valuable metric for quantifying the overall impact of tropical cyclones on the Earth's climate, classifying this index as a duration-based integral of a time series.

The ACE definition given by [28, 29] was adapted to use the monthly output from models, setting a related ACE:

$$ACE = 10^{-1} V_{max}{}^{2} \qquad (1)$$

where V_{max} applied to this work was the monthly mean of maximum daily wind speed in knots, with ACE units being $10^{-1}knots^{2}$.

The primary energy source for TC is the heat from the evaporation that comes from the warmed ocean surface; several studies showed the correlation between sea surface temperature and TC [21, 23, 30, 31]. Additionally, the increase in precipitation rates is largely based on the Clausius-Clapeyron ratio, which produces about a 7% increase in water vapor in the atmosphere by 1°C warming [32, 33]. Thus, the maximum near-surface air temperature at 2 m (TASMAX) expresses a direct physical relationship with the TC occurrence, used here to be an auxiliary proxy to help the discussion ahead.

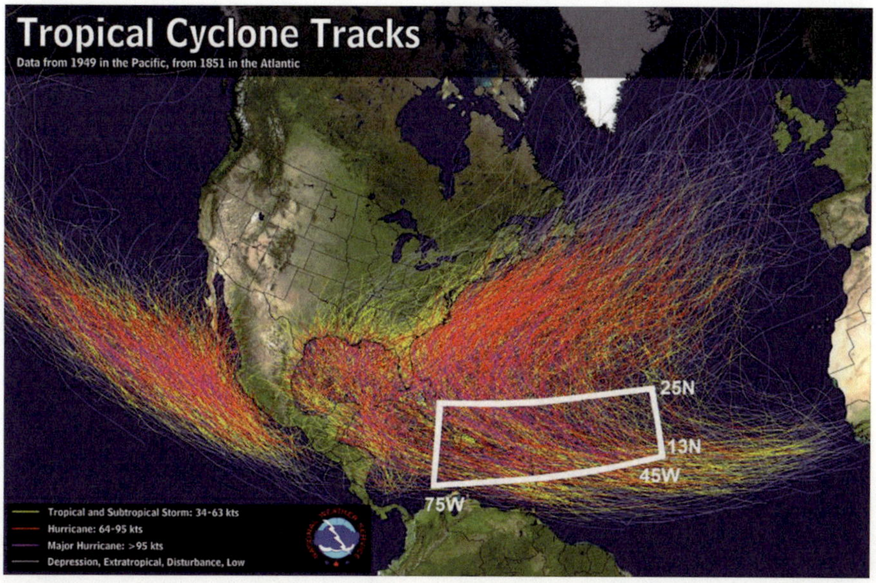

Figure 1.
Tropical cyclone tracks map (adapted from [34]) with the region delimitation for this study: 75 W to 45 W and 13 N to 25 N.

Most tropical cyclones are formed in the intertropical convergence zone (ITCZ). Tropical waves are another important source of atmospheric instability, contributing to the development of about 85% of cyclones over the Atlantic Ocean [35, 36]. TC rarely forms or moves around 5° from the equator where the Coriolis effect is more weak, with most of them appearing between 10 and 30° latitude away from the equator [37]. Thus, the delimited area in the central region in **Figure 1** was chosen as representative to develop the objective of this work.

3. Climate data overview

Climate models have been used to understand how the climate has changed in the past and may change in the future. These models simulate the physics, chemistry, and biology of the atmosphere, land, and oceans, now called Earth system models, and require supercomputers to generate their climate projections.

A set of standard experiments was designed for CMIP, allowing results to be comparable across different model simulations, to see where models agree and disagree on past and future scenarios [38].

CMIP6 historical experiment covers the period 1850–2014, forced by datasets that are largely based on observations, used as an important benchmark for assessing performance through evaluation against observations, and are well suited for quantifying and understanding important climate change response characteristics [38, 27]. The characteristics and forcings included in historical were described in [27]:

- Emissions of short-lived species and long-lived GHGs
- GHG concentrations
- Global gridded land use forcing datasets
- Solar forcing
- Stratospheric aerosol dataset (volcanoes)
- AMIP sea surface temperatures and sea ice concentrations (SICs)
- For simulations with prescribed aerosols, a new approach to prescribe aerosols in terms of optical properties and fractional change in cloud droplet effective radius to provide a more consistent representation of aerosol forcing
- For models without ozone chemistry, time-varying gridded ozone concentrations and nitrogen deposition

Shared socioeconomic pathway (SSP) scenarios are part of a framework designed to span a range of futures in terms of the socioeconomic challenges that they imply for mitigating and adapting to climate change. In short they are:

SSP1 - Low challenges to mitigation and adaptation.
SSP2 - Intermediate challenges to adaptation and mitigation.
SSP3 - High challenges to mitigation and adaptation.
SSP4 - Low challenges to mitigation and high challenges to adaptation.
SSP5 - High challenges to mitigation and low challenges to adaptation [39, 40].

SSP585 results of a complementary effort by SSP narrative and the Representative Concentration Pathways (RCPs), representing the high end of the range of future pathways. SSP5 was chosen for its forcing pathway because its emissions pathway is high enough to produce a radiative forcing of 8.5 W/m^2 by the end of the century, updating RCP8.5 [39, 40]. **Figure 2** summarizes all the current SSP scenarios for CMIP6 in terms of radiative forcing.

3.1 NCEP-DOE AMIP-II reanalysis

Climate reanalysis aims to assimilate historical observational data with numerical models to generate consistent time series of multiple climate variables. These are a comprehensive description of the observed climate as it has evolved during recent decades, providing global datasets at sub-daily intervals, turning possible more detailed approaches, and then having just observation data [41, 42].

NCEP-DOE Reanalysis 2 project performs data assimilation using past data from 1979 through the present. The data is available at PSD portal (https://www.esrl.noaa.gov/psd/data/gridded/data.ncep.reanalysis2.html) in its original four times daily format and as daily averages. The horizontal resolution is 210 km and 28 vertical levels [42].

The zonal and meridional wind components at 2 m and 6-6 hs data were used to compute monthly maximum wind speed. The ACE index was obtained by applying this monthly maximum wind speed in Eq. 1. Similarly, the monthly maximum temperature (TASMAX) was calculated through the daily maximums obtained with 6-6hs data. These variables provided by NCEP reanalysis were used as a reference to evaluate the recent past simulations.

3.2 CMIP6 historical and SSP585 simulations

CMIP6 simulation outputs are available in the Earth System Grid Federation (ESGF), through a distributed data archive developed. The data are hosted on a collection of nodes across the world by modeling centers [43]. The main portal to access the datasets is https://esgf-node.llnl.gov/search/cmip6/.

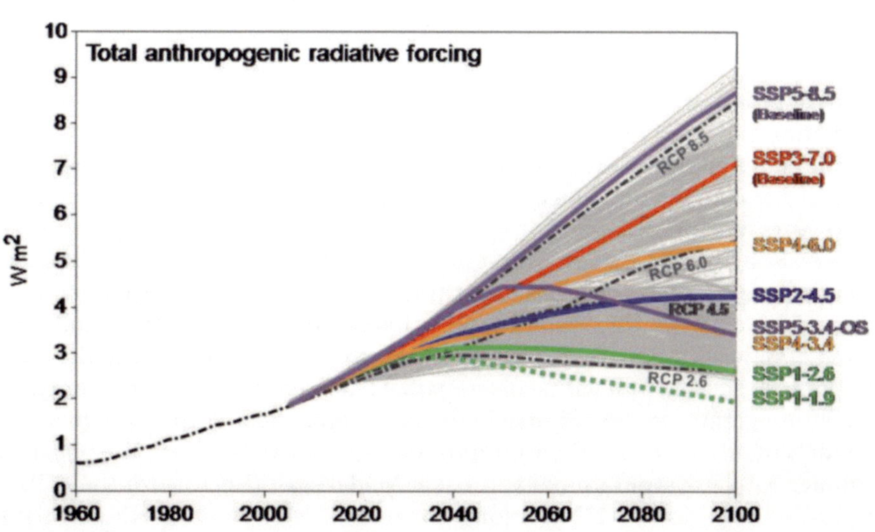

Figure 2.
Anthropogenic radiative forcing for the twenty-first-century scenarios in the ScenarioMIP design (from [39], shown in [40]).

Model	Run	Nominal resolution	Vertical levels	Components
CNRM-CM6-1	r1i1p1f2	250 km	91	AOGCM
CNRM-ESM2-1	r1i1p1f2	250 km	91	AOGCM/BGC/AER/CHEM
IPSL-CM6A-LR	r1i1p1f1	250 km	79	AOGCM/BGC
MPI-ESM1-2-HR	r1i1p1f1	100 km	95	AOGCM
	Institution/Center			**Reference**
CNRM-CM6-1	CNRM (Centre National de Recherches Meteorologiques, Toulouse 31,057, France), CERFACS (Centre Europeen de Recherche et de Formation Avancee en Calcul Scientifique, Toulouse 31,057, France)			[44]
CNRM-ESM2-1				[45]
IPSL-CM6A-LR	Institut Pierre Simon Laplace, Paris 75,252, France			[46]
MPI-ESM1-2-HR	Max Planck Institute for Meteorology, Hamburg 20,146, Germany			[47]

Table 1.
CMIP6 global models and their physical and numerical characteristics.

The complexity of the models, experiments, and methodologies makes it hard for modeling centers to complete the entire archive to participate in CMIP6. Thus, at this time the available datasets to use in this work, for both historical and SSP585, are listed in **Table 1**.

4. Analysis

The models simulate their own climate, with no obligation to get it right exactly when specific events have occurred in relation to observational data. On the other hand, they should be able to represent global or large-scale phenomena such as El Niño, La Niña, ITCZ, and ocean circulation. Thus, the models are expected to represent the average climate of the recent past, as well as to simulate the future in the same direction.

Thus, the regional annual cycle for variables with approximately linear behavior, such as temperature, should be easier to represent. Episodic variables such as precipitation and local wind speed are more difficult to model numerically, given the randomness of events. But it is expected that for long periods, good results will be obtained from the models on average terms, as suggested by the WMO to use at least 30 years for climate studies.

The ACE, because it depends directly on the wind, can be assumed to present results that are less well behaved concerning the reference data than the temperature. This occurs in the results obtained here through the CMIP6 models; the ACE index shows similarities for the monthly climate average through the annual cycle (**Figure 3**), although with a discrepancy between the models higher than the maximum temperature (**Figure 7**). In addition to the nonlinearity involved, the models themselves have their limitations, which may be due to the physical, numerical, or computational approach. Model scaling errors for the ACE index are in the order of −12%. Among the models, MPI-ESM1-2-HR performed better in representing the annual ACE cycle, with a good approximation of the mean monthly values compared to reanalysis, with a correlation of 0.93 and bias error − 1.28%.

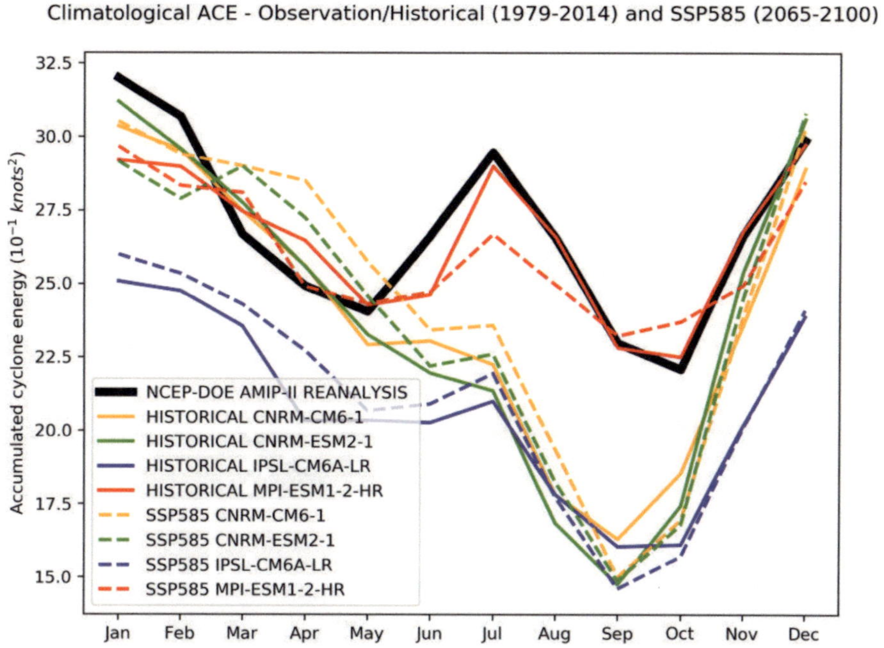

Figure 3.
Annual cycle of the study region for ACE.

The French model IPSL-CM6A-LR has the highest percentage error among the others for ACE, at 22.70% of the reanalysis (**Figure 3**). On the other hand, this same model obtained a better representation of seasonal variability (0.85 correlation) than the two CNRM models, which presented smaller errors (−11%) but with lower correlations, 0.76 (CNRM-CM6-1) and 0.74 (CNRM-ESM2-1). The critical value of the sample correlation, for 95% significance (n-2 degrees of freedom), is 0.576, with all model results performing significant correlations.

The variation coefficient (VC), defined as the ratio of standard deviation by the mean, represents the relative standard deviation, used here to assess whether the models have significant monthly interannual variability or whether they represent climate more closely than stationarity.

The months with the highest percentage variation range from December to May (**Figure 4**), where there is a relative skill of the models, VC values not exceeding 5% from NCEP-DOE Reanalysis 2. In the months from June to November, models have more difficulties to simulate the maximum wind speeds, possibly resulting from the higher activity of the ITCZ in the region selected for the study and being also the months with the high temperatures of the year (**Figure 7**). The MPI-ESM1-2-HR model best quantified the interannual ACE variations for the months with the highest CT activity, followed by the IPSL model, erring only in magnitude, hitting the temporal evolution in most months.

The polynomial curve fitting creates an approximating function that attempts to capture important patterns in the data while leaving out noise or other fine-scale structures/rapid phenomena. This method can aid in data analysis by being able to extract more information from the data as long as the assumption of smoothing is reasonable and to provide analysis that is both flexible and robust.

The first-degree coefficient represents the linear trend of the data, and, as shown in **Figure 5**, the NCEP reanalysis has a small negative trend in annual ACE over the recent past (1979–2014). With the same trend signal, IPSL-CM6A-LR follows the observation pathway, while the other three models simulate a positive

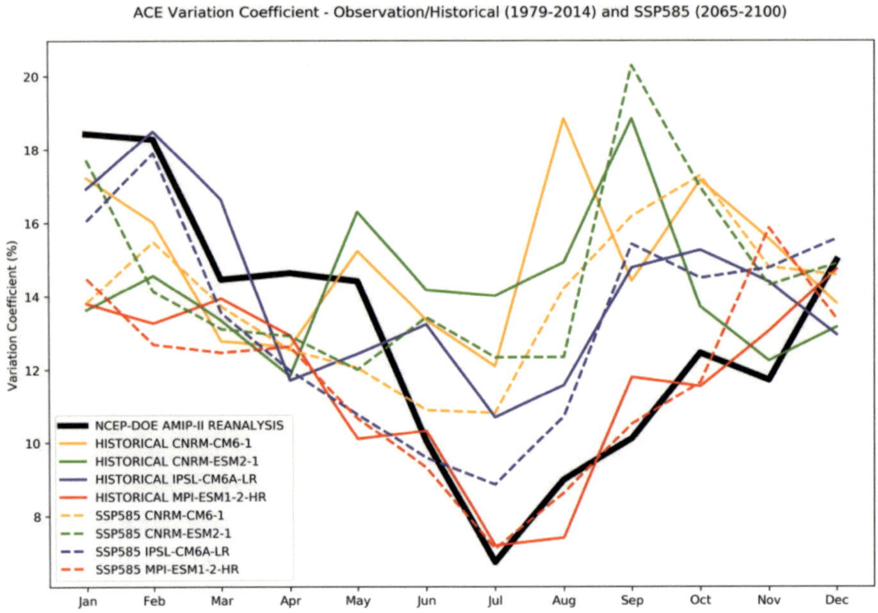

Figure 4.
Percentage variation coefficient for ACE monthly values.

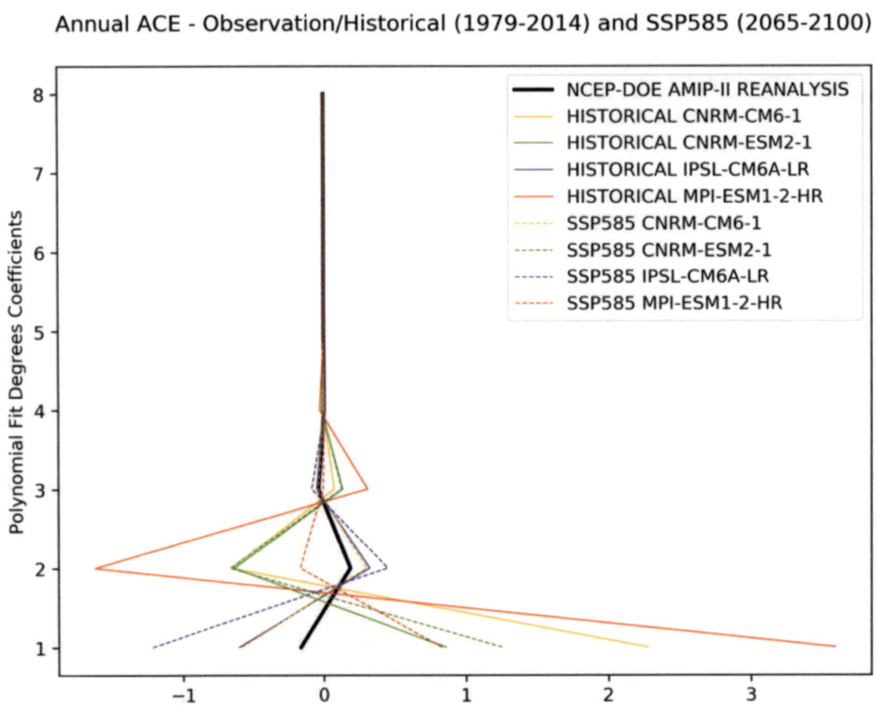

Figure 5.
Polynomial adjustment coefficients for annual ACE of the study region.

trend. From the second-degree coefficient to further ahead, the adjustments are related to patterns with more oscillatory rates, and in the present analysis, this type of signal has no significance. Thus, it can be assumed that models with coefficient values close to reanalysis, in modulus, should have a similar pattern of variability in different modes. The German model was the most difficult to obtain the adjustment, probably because it has a higher horizontal resolution, making it possible to discretize more climate phenomena, which has coefficient values more distant than the obtained for the reanalysis.

The projection of annual ACE for the twenty-first century (**Figure 6**) has a similar average behavior among models, without abrupt trend changes, presenting modes of variation not far from the simulated for the recent past. The long-term trend for the period 2065–2100 is an increase in the average annual ACE values for the CNRM-ESM2-1 and MPI-ESM1-2-HR models and a reduction for the IPSL-CM6A-LR and CNRM-CM6-1, with no majority agreement.

The TASMAX annual cycle has a good performance by the models; in terms of seasonality, all models show suitable patterns, with low errors in representing the evolution of the monthly cycle. The bias error is a problematic aspect, the three French models have sub estimate ~2°C, while MPI-ESM1-2-HR fits almost the entire NCEP reanalysis climatology (**Figure 7**).

The annual TASMAX projections for the future (**Figure 8**) are similar to that described in the IPCC Special Report on the impacts of global warming of 1.5°C above preindustrial levels and related global greenhouse gas emission pathways [48], in which there is a high confidence that the estimated anthropogenic global warming is currently increasing at 0.2°C per decade due to past and ongoing emissions.

The mid- and long-term ACE future projections for most models analyzed indicate to the increase of the index and just the MPI-ESM1-2-HR follows a different pathway (**Figure 9**). The approaches used in the results shown in **Figures 9** and **10** consist of calculating the future percentage change over the periods and applying this change to the reanalysis recent past value. This way, the projection has no bias error associated to it, bringing the right value expected in the future for the projection.

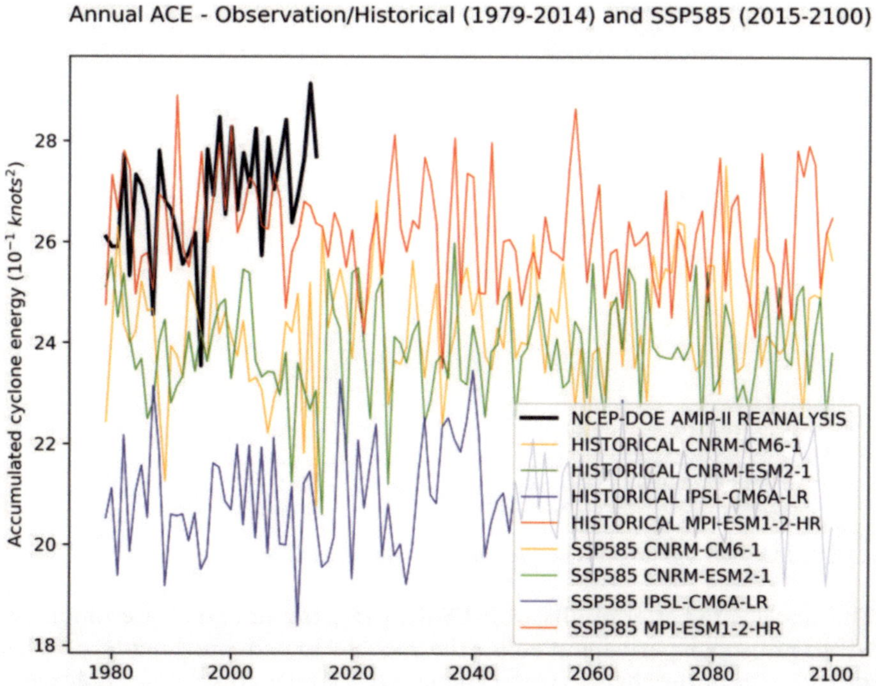

Figure 6.
Annual ACE time series for recent past and future simulation under SSP585.

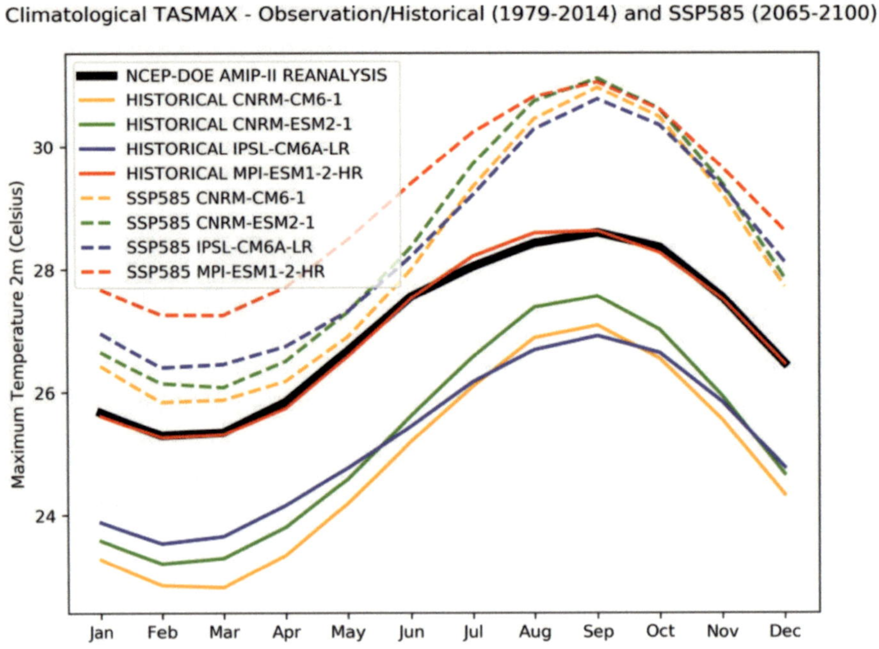

Figure 7.
Annual cycle of mean over the study region for TASMAX.

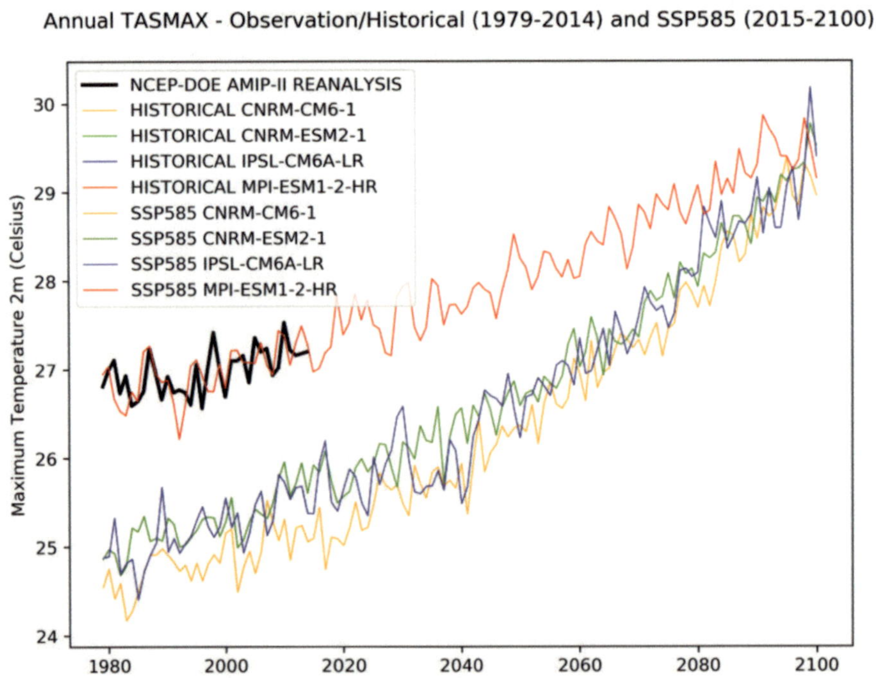

Figure 8.
Annual TASMAX time series for recent past and future simulation under SSP585.

The model with better results, MPI-ESM1-2-HR, trends to increase annual ACE under the projection period, but points the opposite to mid- and long-term mean (**Figure 9**). One of the changes in the annual cycle is an increase in the index in months where TC activity is not intense, as in the months of the beginning and end of the year, in which there is also an increase in VC. These factors suggest that the

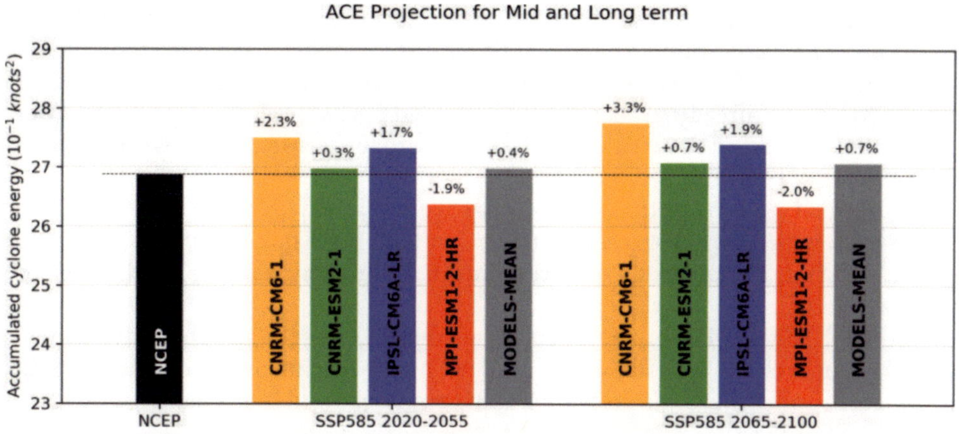

Figure 9.
Future ACE projection under SSP585 for mid (2020–2055) and long (2065–2100) terms.

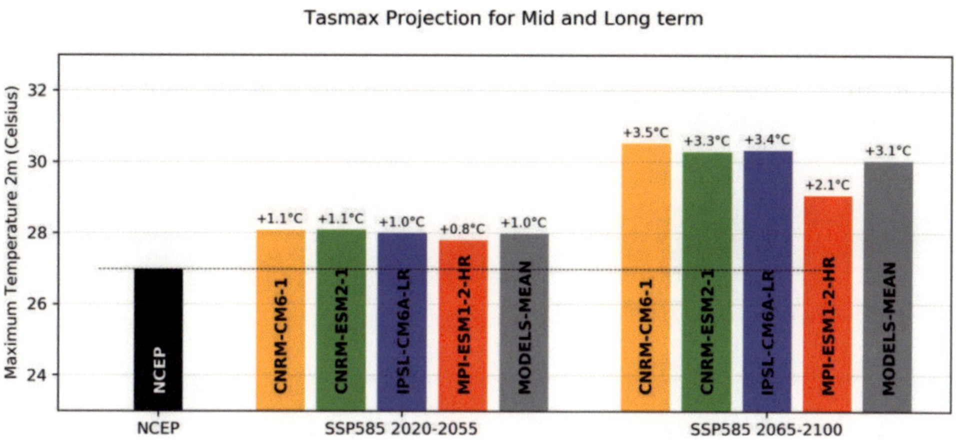

Figure 10.
Future TASMAX projection under SSP585 for mid (2020–2055) and long (2065–2100) terms.

seasons and their interannual variations in cyclonic activity will be affected by the forcing on the climate system, in this case, under the scenario of high GHG emissions and high challenges to mitigation SSP585.

The MODELS-MEAN projection (**Figures 9** and **10**) was computed by the weight mean, considering the annual cycle correlation value as the weight for each model. Thus, MODELS-MEAN performs a more confident projection. The results for that concern to a future with more chances of facing more tropical cyclone activity, plus the huge long-term TASMAX increase of 3.1°C (**Figure 10**); the twenty-first century may experiment more heavy cyclones and stronger storms with more frequency, as indicated by other studies [21, 23, 25, 26].

5. Conclusions

The accumulated cyclone energy index adapted for this work has made it simpler to assess the recent past and to obtain projections of CMIP6 models, given the use of monthly data directly.

In the present climate evaluation (1979–2014), reasonable results were obtained for the ACE index; the French models of lower horizontal and vertical resolution showed more difficulties to represent the index, while the Max Planck Institute model demonstrated ability to simulate the climate with more accuracy than the others, presenting values of both ACE and TASMAX very close to NCEP Reanalysis 2.

TASMAX was already expected to obtain good results numerically; in terms of seasonality all models show suitable patterns, with low errors in representing the evolution of the monthly annual cycle.

The annual ACE projection has a similar average behavior among models in the recent past, without abrupt trend changes, but with no major agreement to increase or reduce trend. The mid- and long-term mean for most models analyzed shows an increase in ACE.

The MPI-ESM1-2-HR projections suggest that the seasons and their interannual variations in cyclonic activity will be affected by the forcing on the climate system, in this case, under the scenario of high GHG emissions and high challenges to mitigation SSP585.

The results indicate to a future with more chances of facing more tropical cyclone activity, plus the mean increase of 3.1°C in maximum daily temperatures, and more heavy cyclones and stronger storms with more frequency may be experimented, as indicated by other studies [21, 23, 25, 26].

The study needs to be expanded, including more models, to increase the range of results and to narrow down potential trends that may occur in ensemble analysis.

Author details

Sullyandro Oliveira Guimarães
Federal University of Ceará, Fortaleza, Ceará, Brazil

*Address all correspondence to: sullyandro@gmail.com

References

[1] Bindoff NL, Stott PA, Achuta Rao KM, Allen MR, Gillett N, Gutzler D, et al. Detection and attribution of climate change: From global to regional. In: Stocker TF, Qin D, Plattner G-K, Tignor M, et al., editors. Climate Change 2013: The Physical Science Basis. Contribution of Working Group I to the Fifth Assessment Report of the Intergovernmental Panel on Climate Change. Cambridge, United Kingdom/New York, NY, USA: Cambridge University Press; 2013. pp. 867-952

[2] Myhre G, Shindell D, Bréon F-M, Collins W, Fuglestvedt J, Huang J, et al. Anthropogenic and natural radiative forcing. In: Stocker TF, Qin D, Plattner G-K, Tignor M, et al., editors. Climate Change 2013: The Physical Science Basis. Contribution of Working Group I to the Fifth Assessment Report of the Intergovernmental Panel on Climate Change. Cambridge, United Kingdom/New York, NY, USA: Cambridge University Press; 2013. pp. 659-740

[3] PAGES 2K Consortium. Continental-scale temperature variability during the past two millennia. Nature Geoscience. 2013;**6**(5):339-346. DOI: 10.1038/ngeo1797

[4] Marcott SA, Shakun JD, Clark PU, Mix AC. A reconstruction of regional and global temperature for the past 11,300 years. Science. 2013;**339**(6124):1198-1201. DOI: 10.1126/science.1228026

[5] Cheng L, Trenberth KE, Fasullo J, Boyer T, Abraham J, Zhu J. Improved estimates of ocean heat content from 1960 to 2015. Science Advances. 2017;**3**(3):e1601545. DOI: 10.1126/sciadv.1601545

[6] Church JA, White NJ, Konikow LF, Domingues CM, Cogley JG, Rignot E, et al. Revisiting the Earth's sea-level and energy budgets from 1961 to 2008. Geophysical Research Letters. 2011;**38**(18):L18601. DOI: 10.1029/2011GL048794

[7] Fahey DW, Doherty S, Hibbard KA, Romanou A, Taylor PC. Physical drivers of climate change. Climate science special report: Fourth national climate assessment, Vol. I. In: Wuebbles DJ, Fahey DW, Hibbard KA, Dokken DJ, Stewart BC, Maycock TK, editors. U.S. Global Change Research Program. Washington, DC, USA; 2017. pp. 73-113. DOI: 10.7930/J0513WCR

[8] Wuebbles DJ, Easterling DR, Hayhoe K, Knutson T, Kopp RE, Kossin JP, et al. Our globally changing climate. Climate science special report: Fourth National Climate Assessment, volume I. In: Wuebbles DJ, Fahey DW, Hibbard KA, Dokken DJ, Stewart BC, Maycock TK, editors. U.S. Global Change Research Program. Washington, DC, USA; 2017. pp. 35-72. DOI: 10.7930/J08S4N35

[9] Anderson BT, Knight JR, Ringer MA, Yoon J-H, Cherchi A. Testing for the possible influence of unknown climate forcings upon global temperature increases from 1950 to 2000. Journal of Climate. 2012;**25**(20):7163-7172. DOI: 10.1175/jcli-d-11-00645.1

[10] Bhatia KT, Vecchi GA, Knutson TR, et al. Recent increases in tropical cyclone intensification rates. Nature Communications. 2019;**10**:635. DOI: 10.1038/s41467-019-08471-z

[11] Hayhoe K, Wuebbles DJ, Easterling DR, Fahey DW, Doherty S, Kossin J, et al. Our changing climate. Impacts, risks, and adaptation in the United States: Fourth national climate assessment, volume II. In: Reidmiller DR, Avery CW, Easterling DR, Kunkel KE, Lewis KLM, Maycock TK, Stewart BC, editors. U.S. Global Change Research Program.

"FOURTH NATIONAL CLIMATE ASSESSMENT Volume II: Impacts, Risks, and Adaptation in the United States". 2018. Retrieved: 7 September 2019

[12] Emanuel KA. Increasing destructiveness of tropical cyclones over the past 30 years. Nature. 2005;**436**:686-688

[13] Elsner JB, Kossin JP, Jagger YH. The increasing intensity of the strongest tropical cyclones. Nature. 2008;**455**:92-95

[14] Holland G, Bruyère CL. Recent intense hurricane response to global climate change. Climate Dynamics. 2014;**42**:617-627

[15] Hoyos CD, Agudelo PA, Webster PJ, Curry JA. Deconvolution of the factors contributing to the increase in global hurricane intensity. Science. 2006;**312**:94-97

[16] Kossin JP, Olander TL, Knapp KR. Trend analysis with a new global record of tropical cyclone intensity. Journal of Climate. 2013;**26**:9960-9976

[17] Wang C, Wang X, Weisberg RH, Black ML. Variability of tropical cyclone rapid intensification in the North Atlantic and its relationship with climate variations. Climate Dynamics. 2017;**49**:3627-3645

[18] Balaguru K, Foltz GR, Leung LR. Increasing magnitude of hurricane rapid intensification in the central and eastern tropical Atlantic. Geophysical Research Letters. 2018;**45**:4238-4247

[19] Mei W, Xie SP. Intensification of landfalling typhoons over the Northwest Pacific since the late 1970s. Nature Geoscience. 2016;**9**:753-757

[20] Emanuel KA. Thermodynamic control of hurricane intensity. Nature. 1999;**401**:665

[21] Knutson T, Camargo SJ, JCL C, Emanuel K, Ho C-H, Kossin J, et al. Tropical cyclones and climate change assessment: Part II. Projected response to anthropogenic warming. Bulletin of the American Meteorological Society. 2019. DOI: 10.1175/BAMS-D-18-0194.1. BAMS–D–18-0194.1

[22] Christensen JH, Krishna Kumar K, Aldrian E, An S-I, Cavalcanti IFA, de Castro M, et al. Climate phenomena and their relevance for future regional climate change. In: Stocker TF, Qin D, Plattner G-K, Tignor M, Allen SK, Boschung J, et al., editors. Climate Change 2013: The Physical Science Basis. Contribution of Working Group I to the Fifth Assessment Report of the Intergovernmental Panel on Climate Change. Cambridge, United Kingdom, New York, NY, USA: Cambridge University Press; 2013. 1535 pp

[23] Knutson TR, McBride JL, Chan J, Emanuel K, Holland G, Landsea C, et al. Tropical cyclones and climate change. Nature Geoscience. 2010;**3**:157-163. DOI: 10.1038/ngeo0779

[24] IPCC. In: Stocker TF, Qin D, Plattner G-K, Tignor M, et al., editors. Climate Change 2013: The Physical Science Basis. Contribution of Working Group I to the Fifth Assessment Report of the Intergovernmental Panel on Climate Change. Cambridge, United Kingdom and New York, NY, USA: Cambridge University Press; 2013. 1535 pp

[25] Walsh KJE, Camargo SJ, Vecchi GA, Daloz AS, Elsner J, Emanuel K, et al. Hurricanes and climate: The U.S. CLIVAR working group on hurricanes. Bulletin of the American Meteorological Society. 2015;**96**:997-1017

[26] Matthew RA, Sanders BF, Aghakouchak A, Salvadori G, Moftakhari HR. Compounding effects

of sea level rise and fluvial flooding. Proceedings of the National Academy of Sciences. 2017;**114**(37):9785-9790. DOI: 10.1073/pnas.1620325114

[27] Eyring V, Bony S, Meehl GA, Senior CA, Stevens B, Stouffer RJ, et al. Overview of the coupled model Intercomparison project phase 6 (CMIP6) experimental design and organization. Geoscientific Model Development. 2016;**9**:1937-1958. DOI: 10.5194/gmd-9-1937-2016

[28] Bell GD et al. Climate assessment for 1999. Bulletin of the American Meteorological Society. 2000;**81**:S1-S50

[29] Separating the ACE Hurricane Index into Number, Intensity, and Duration by Carl Drews [Internet]. 2007. Available from: https://acomstaff.acom.ucar.edu/drews/hurricane/SeparatingTheACE.html [Accessed: 13 October 2019]

[30] Lin Y, Zhao M, Zhang M. Tropical cyclone rainfall area controlled by relative sea surface temperature. Nature Communications. 2015;**6**:6591. DOI: 10.1038/ncomms7591

[31] Emanuel K. Environmental factors affecting tropical cyclone power dissipation. Journal of Climate. 2007;**20**:5497-5509

[32] Wright DB, Knutson TR, Smith JA. Regional climate model projections of rainfall from U.S. landfalling tropical cyclones. Climate Dynamics. 2015;**45**(11-12):3365-3379. DOI: 10.1007/s00382-015-2544-y

[33] Knutson TR, Sirutis JJ, Zhao M. Global projections of intense tropical cyclone activity for the late twenty-first century from dynamical downscaling of CMIP5/RCP4.5 scenarios. Journal of Climate. 2015;**28**(18):7203-7224. DOI: 10.1175/JCLI-D-15-0129.1

[34] Tropical Cyclone Climatology [Internet]. 2017. Available from: https://www.nhc.noaa.gov/climo/ [Accessed: 20 November 2019]

[35] Avila LA, Pasch RJ. Atlantic tropical systems of 1993. Monthly Weather Review. 1995;**123**(3):887-896. DOI: 10.1175/1520-0493

[36] Landsea CW. A climatology of intense (or major) Atlantic hurricanes. Monthly Weather Review. 1993;**121**(6):1703-1713. DOI: 10.1175/1520-0493

[37] Dowdy AJ, Qi L, Jones D, Ramsay H, Fawcett R, Kuleshov Y. Tropical cyclone climatology of the South Pacific Ocean and its relationship to El Niño-Southern oscillation. Journal of Climate. 2012;**25**(18):6108-6122. DOI: 10.1175/JCLI-D-11-00647.1

[38] Pascoe C, Lawrence BN, Guilyardi E, Juckes M, Taylor KE. Designing and documenting experiments in CMIP6. Geoscientific Model Development Discussions. 2019. https://doi.org/10.5194/gmd-2019-98

[39] Riahi K, van Vuuren DP, Kriegler E, Edmonds J, O'Neill BC, Fujimori S, et al. The shared socioeconomic pathways and their energy, land use, and greenhouse gas emissions implications: An overview. Global Environmental Change. 2017;**42**:153-168. DOI: https://doi.org/10.1016/j.gloenvcha.2016.05.009

[40] O'Neill BC, Tebaldi C, van Vuuren DP, Eyring V, Friedlingstein P, Hurtt G, et al. The scenario model Intercomparison project (ScenarioMIP) for CMIP6. Geoscientific Model Development. 2016;**9**:3461-3482. DOI: 10.5194/gmd-9-3461-2016

[41] Uppala S et al. The ERA-40 re-analysis. Quarterly Journal of the Royal Meteorological Society. 2005;**131**:2961-3012. DOI: 10.1256/qj.04.176

[42] Kanamitsu M, Ebisuzaki W, Woollen J, Yang S-K, Hnilo JJ, Fiorino M, et al. NCEP-DOE AMIP-II Reanalysis (R-2). Bulletin of the American Meteorological Society. 2002:1631-1643

[43] CMIP6 Guidance for Data Users [Internet]. 2019. Available from: https://pcmdi.llnl.gov/CMIP6/Guide/dataUsers.html [Accessed: 20 November 2019]

[44] Voldoire A, Saint-Martin D, Sénési S, Decharme B, Alias A, Chevallier M, et al. Evaluation of CMIP6 DECK experiments with CNRM-CM6-1. Journal of Advances in Modeling Earth Systems. 2019;**11**:2177-2213. https://doi.org/10.1029/2019MS001683

[45] Séférian R, Nabat P, Michou M, Saint-Martin D, Voldoire A, Colin J, et al. Evaluation of CNRM earth-system model, CNRM-ESM2-1: Role of earth system processes in present-day and future climate. Journal of Advances in Modeling Earth Systems. 2019;**11**:4182-4227. https://doi.org/10.1029/2019MS001791

[46] IPSL Climate Modelling Centre [Internet]. 2019. Available from: https://cmc.ipsl.fr/ipsl-climate-models/ipsl-cm6/ [Accessed: 11 November 2019]

[47] Mauritsen T et al. MPI-ESM: Developments in the MPI-M earth system model version 1.2 (MPI-ESM1.2) and its response to increasing CO_2. Journal of Advances in Modeling Earth Systems. 2019;**11**:998-1038. DOI: 10.1029/2018MS001400

[48] IPCC. Summary for policymakers. In: Masson-Delmotte V, Zhai P, Pörtner H-O, Roberts D, Skea J, Shukla PR, et al., editors. Global Warming of 1.5°C. An IPCC Special Report on the Impacts of Global Warming of 1.5°C above Pre-industrial Levels and Related Global Greenhouse Gas Emission Pathways, in the Context of Strengthening the Global Response to the Threat of Climate Change, Sustainable Development, and Efforts to Eradicate Poverty. Geneva, Switzerland: World Meteorological Organization; 2018. 32 pp